A PRIMER ON ENVIRONMENTAL SCIENCES

Matthew N. O. Sadiku
Uwakwe C. Chukwu
Olaniyi D. Olaleye

AuthorHouse™
1663 Liberty Drive
Bloomington, IN 47403
www.authorhouse.com
Phone: 833-262-8899

This book is printed on acid-free paper.

ISBN: 978-1-6655-4754-3 (sc)
978-1-6655-6278-2 (hc)
978-1-6655-4755-0 (e)

Library of Congress Control Number: 2021925439

Print information available on the last page.

Published by AuthorHouse 03/10/2023

authorHOUSE®

A PRIMER ON ENVIRONMENTAL SCIENCES

Table of Contents

CHAPTER 1

INTRODUCTION

"If you want to learn about the health of a population, look at the air they breath, the water they drink, and the places where they live." –Hippocrates,

1.1 INTRODUCTION

The environment is an important part of our daily lives. The environment sustains life. Man needs to know the importance of the environment and keeps the environment as healthy as possible. Environmentalists around the world are constantly seeking sustainable solutions to restore a sustainable environment.

Today, we are facing some serious environmental challenges. The environmental crisis is one of the biggest challenges of the 21st century. For this reason, environmental problems are now much-discussed around the globe. Environmental problems or threats include [1]: (1) Climate change, which is a change in earth's climate; (2) Rapid urbanization and industrialization have destroyed a substantial part of natural vegetation and forced many wild animals on the verge of extinction; (3) Overconsumption of goods, which encourages the acquisition of goods in ever increasing amounts; (4) Ozone depletion, which is the wearing out or reduction of the amount of ozone in the stratosphere; (5) Deforestation, which is the cutting down of trees and the destruction of natural vegetation; (6) Desertification, which is the degradation of land in arid, semi-arid, and dry-sub-humid areas..

It is essential to make the public aware of the dire consequences of the environmental degradation [2]. Figure 1.1 shows how the environment impact our health [3]. Environmental science is, therefore, an interdisciplinary study of how the earth works and how we can deal with the environmental issues we face.

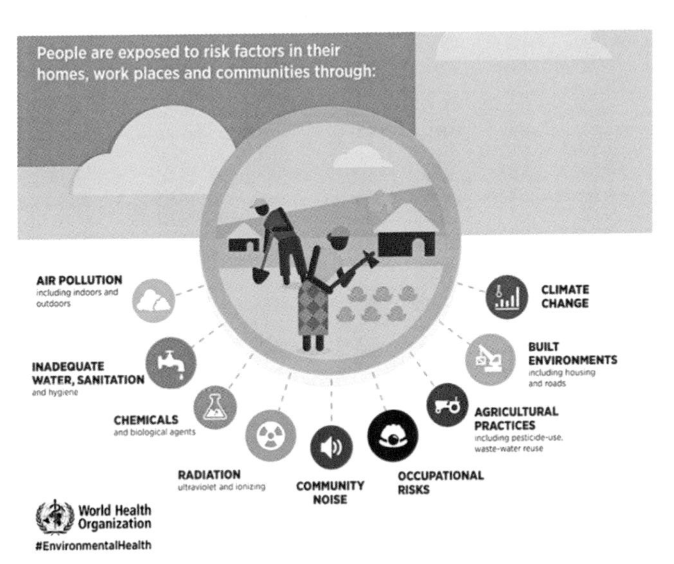

Figure 1.1 How the environment impact our health [3].

The threat to the environment has been increasing. People around the world are experiencing environmental problems such as rising population, poverty, air and water; pollution, food insecurity, flood management, land management, endangered species, etc. Careless handling of today's environment would affect tomorrow's generation. Hence, a judicious use of our resources is called for.

This chapter provides an introduction to environmental science. It begins by describing different types of environment. It presents what environmental science is all about. It also covers different components of environmental science. It highlights some challenges faced by environmental scientists. Finally, it concludes with some comments.

1.2 TYPES OF ENVIRONMENT

The word "environment" means to "surround, enclose, or encircle." Environment refers to surroundings in which living beings live and non-living things exist. Environment belongs to all the living beings. Importance of the environment was realized in the 1960s and reached its climax in 1970, with the celebration of "Earth Day" under the auspices of the United Nation. Environment plays a major role in the healthy living of human beings. As shown in Figure 1.2, environment includes air, water, land, living organisms, solar energy, and materials surrounding us [4].

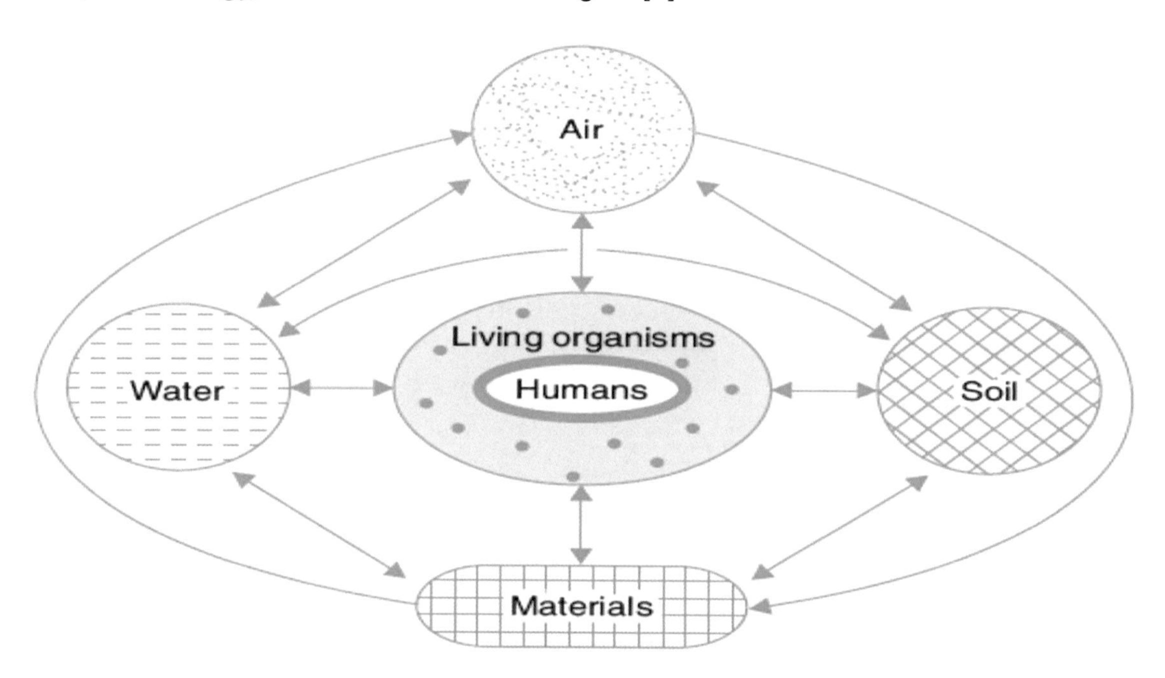

Figure 1.2 The concept of environment [4].

Our existence on this planet requires abundance of land, air, and water. Without doubt, the rapid technological development in the twentieth century has very adversely impacted the environment. To ensure a sustainable development, we need to know something about how our environment works. Environment may be regarded as all external conditions and factors that affect living organisms. It is generally equated with nature wherein physical components of the planet earth such as earth, air, and water-support and affect life in the biosphere [5]. Our environment affects everything we do— from climate changes to air quality, and much more.

Since environment is a combination of physical and biological factors, it contains both living and non-living components. Everyone may be affected by environmental issues like global warming, ozone layer depletion, energy resources, acid rain, nuclear accidents, loss of global biodiversity, etc. Some environmentalists consider the following diversity of environments: Coastal environment, plateau environment, mountain environment, lake environment, river environment, maritime environment, cultural environment, built or man-made environment, living or biotic environment, smart environment, green environment, etc. These environments can be classified as discussed in the following sections.

1.1.1 Physical or Living Environment

Environment can be divided into physical or abiotic and living or biotic environment. Physical environment consists of solid, liquid, and gas. These three elements signify lithosphere, hydrosphere, and atmosphere respectively. Living environment consists of plants (flora) and animals (fauna), including human beings. The physical elements of the planet earth, such as terrain, soil, water, climate, flora, and fauna formed man's environment.

1.1.2 Built Environment

This is a man-made environment that provides the setting for human activity, ranging from buildings to cities. This is when the natural environment is deliberately controlled by mankind. The built environment is a material, spatial, and cultural product of human labor that combines physical elements and energy in forms for living, working, and playing. Examples include homes, building, offices, aquariums, cities, community parks, and laboratories. The built environment has a strong impact on the people who use it; it affects both their human and environmental health.

1.2.3 Smart Environments

The word "smart" means intelligent, while the word "environment" means our surrounding. The term "smart environment" refers to a living environment that, by its connected sensors and actuators, is capable of providing intelligent and contextualized support to its inhabitants. Smart environments are essentially physical worlds interwoven with sensors, actuators, displays, and computational devices, embedded into everyday objects and connected through a communication network. They are sensitive to the needs of occupants, can anticipate their behavior, and respond accordingly. Their goal is to improve the experience of inhabitants of the environment.

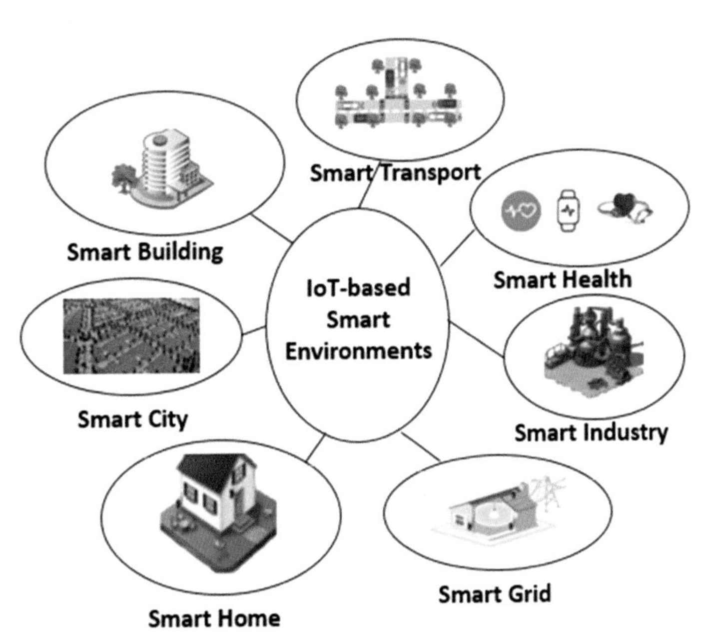

Figure 1.3 Examples of IoT-based smart environments [6].

As illustrated in Figure 1.3, smart environment is used differently to refer to smart rooms, smart houses, smart buildings, smart cities, smart grids, smart offices, smart healthcare centers, smart classrooms, smart factories, smart workplaces, smart agriculture, kindergartens, car, and smart laboratories [6]. A smart environment is one that is able to acquire and apply knowledge about the environment and its inhabitants in order to improve their experience in that environment. It must be able to determine and predict the environment and the characteristic response behavior of the inhabitants. The predictive feature relies on tools from artificial intelligence (the theory and development of computer systems that is capable of performing tasks that normally require human intelligence).

The concept of smart environments evolves from ubiquitous computing, which is a technology in which invisible computers are embedded and connected with all things enabling the use of computers anywhere at any time. A smart environment (also called ambient intelligence) is a physical space where different kinds of smart devices are working together to make inhabitants live more comfortably. Its goal is to acquire and exploit knowledge of the environment so as to adapt itself to its inhabitants' requirements [8].

1.1.3 Green Environment

This refers to the concerns for environmental conservation. This is demonstrated in supporting practices such as conservation practices, investment in renewable energy, and taking actions to stop climate change, and reduce carbon footprint. Being environmentally friendly will affect corporate policy and individual behavior. Many individuals cultivate personal gardens and such activity can have direct health benefits. Together, we can leave a greener footprint on the earth for the benefit of the next generation [9].

Due to increasing concern about health and the environment, environment friendly practices have been introduced everywhere. Becoming more environmentally friendly, is known as "going green." Going green is practicing an environmentally-mindful lifestyle that contributes towards protecting the environment and conservation of the natural resources. It also means embracing a way of life that helps preserve the environment by reducing, reusing, and recycling items. There is dire need to go green in order to save the environment and ourselves.

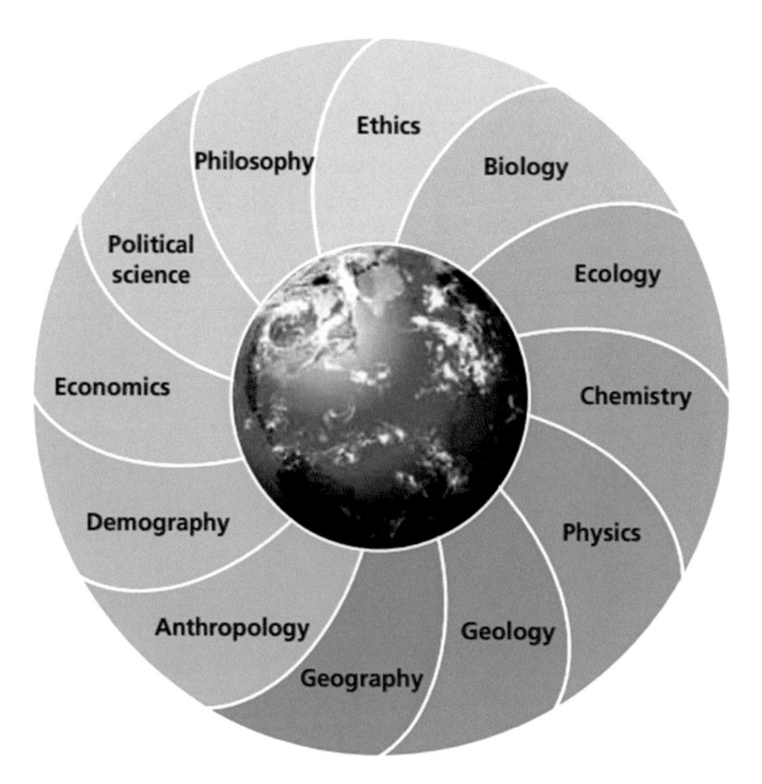

Figure 1.4 Environmental science is an interdisciplinary field [10].

1.2 WHAT IS ENVIRONMENTAL SCIENCE?

Science is the desire to find out how something happens. It may be regarded as all of the fields of study that attempt to comprehend the nature of the universe. Environmental science is an interdisciplinary field that applies principles from all the known technology and sciences to study the environment and provide solutions to environmental problems. It is the science of physical phenomena in the environment. As shown in Figure 1.4, environment science is a multi-disciplinary discipline that embraces biology, chemistry, physics, agriculture, public health, technology, law, arts, statistics, microbiology, biochemistry, geology, economics, law, sociology, arts, geography, resource management, etc. [10]. It studies all aspects of the environment in an interdisciplinary manner. Nearly any topic can be studied in terms of its relationship with the natural environment. The field of environmental science is involved with natural science, technical science, social science, and administrant science.

As shown in Figure 1.5, environmental scientists tend to understand the interactions between humans and the natural world [11]. They employ a systems approach to the analysis of complex environmental problems, which often include an interaction of physical, chemical, and biological processes [12]. Their environmental efforts address global issues. There are several environmental thinkers including Charles Darwin, Ralph Emerson, Henry Thoreau, John Muir, Aldo Leopald, and Rachel Carson. In the US, the National Environmental Policy Act (NEPA) of 1969 set forth requirements for analysis of federal government actions on specific environmental issues.

Figure 1.5 Environmental scientists understand the interactions between humans and the natural world [11].

1.4 COMPONENTS OF ENVIRONMENTAL SCIENCE

Environmental science includes the following components [2,13].

- *Ecology:* Although the terms "environmental science" and "ecology" are often used interchangeably, ecology should be considered a subset of environmental science. For this reason, ecology is often referred to as environmental biology. Ecology is the study of the relationships between living organisms and their physical environment. It is the science of the relations of all organisms to all their environments.

- *Environmental Chemistry*: This is the study of chemical alterations in the environment. It is divided into three main classifications: Abiotic, biotic, and energy components in chemical science. Therefore, it is an interdisciplinary field that studies the presence and impact of chemicals in air, soil, water, and living environment. It deals with the impact of chemicals and their effects on human health and organisms in the environment. It helps us trace and control contaminants. It involves understanding the chemical processes that occur in water, air, terrestrial, and living environments. Main areas of study include soil contamination, water pollution, chemical degradation in the environment, etc.
- *Environmental Engineering:* Traditionally, environmental engineering started as part of civil engineering. Now, it is one of the fastest growing and most complex disciplines of engineering. Environmental engineering applies science and engineering principles to develop ways to protect human health and minimize the adverse effects of human activities on the environment. The field emerged in response to widespread public concern about environmental degradation such water and air pollution. Environmental engineers solve problems, design systems, and provide solutions to various environmental problems [14].
- *Environmental Economics*: This is a branch of economics that focuses on environmental problems of pollution of earth, air, and water. It studies the financial impact of environmental policies and the effects of environmental policies on the economy. It also deals with the impact of economic activities on the environment. It supports environmental policies to deal with air pollution, water quality, toxic substances, solid waste, and global warming. It considers issues such as the conservation and valuation of natural resources, pollution control, waste management, and recycling [15].
- *Environmental Management:* This is a multidisciplinary area that is concerned with the management of human activities and their impacts on the natural environment. It is basically about making decisions on the use of natural resources. It involves pressing issues of justice and survival. Environmental managers consist of a diverse group of people including academics, policy-makers, government, non-governmental organization (NGO) workers, company employees, civil servants, and other individuals or groups who desire to control the direction and pace of development [16].
- *Environmental Laws:* These laws deal with issues related to environmental challenges and conservation of natural resources. Environmental laws have been developed in response to growing concern over issues impacting the environment worldwide. They are legal enactments designed to consciously preserve the environment or protect the environment from damage. They are meant to protect human health as well as the environment. The laws cover pollutants (like pollution of water, air and soil related to greenhouse gas emissions, waste disposals, and acid rain), natural resource conservation (like hunting of endangered species, deforestation, and depletion of natural resources), and energy issues (like global warming and climate change).
- *Biodiversity*: This is also known as "biological diversity." It refers to the existence of a number of different species of plants and animals in an environment. Diversity is the number of species found in a given community. Biodiversity refers to the species richness of an area. It is substantially greater in some areas than in the others. Biodiversity is at local, national, and global levels. Biodiversity is diminished or destroyed in a number of ways either by natural changes or by human disruption. As species become extinct, the balance of nature is disturbed and redistributed to great extent. In view of the degree of threat to biodiversity around the world, there is an urgent need to conserve biodiversity in the world.

- *Natural Resources:* These are the resources we obtain from nature or earth. They occur naturally and humans cannot make them. Natural resources can be classified based on their origin, level of development and uses, stock or deposits, and their distribution. Natural resources can be classified as renewable or non-renewable. We also have water resources, mineral resources, land resources, food resources, energy resources, and forest resources. Environmental laws propose regulations that promote the conservation of natural resources. Energy crisis is a result of many different strains on our natural resources (for instance, strain on fossil fuels such as oil, gas, and coal due to overconsumption can put a strain on our water and oxygen resources by causing pollution).
- *Energy Resources:* Energy and environment mark the two key challenges to human beings for the future sustainability of the planet. Energy is simply the capacity to do work. Energy appears in several forms. The sun is our primary source of energy. Other sources of energy include water, fossil fuels, petroleum products, and nuclear power plants. Modern lifestyle, industrialization, urbanization, and increase in population have increased global energy requirement. There are two types of energy: renewable energy or non-renewable energy. Renewable energy (such as hydropower, solar, wind, and geothermal) uses resources that are constantly replaced and are usually less polluting. Non-renewable energy (such as fossil fuels, coal, petroleum, and natural gas) cannot be replenished after their depletion. Energy crisis rises when there is any significant bottleneck in the supply of energy resources to an economy. Overpopulation, for example, can cause a significant drain on our energy resources thereby resulting in energy crisis.

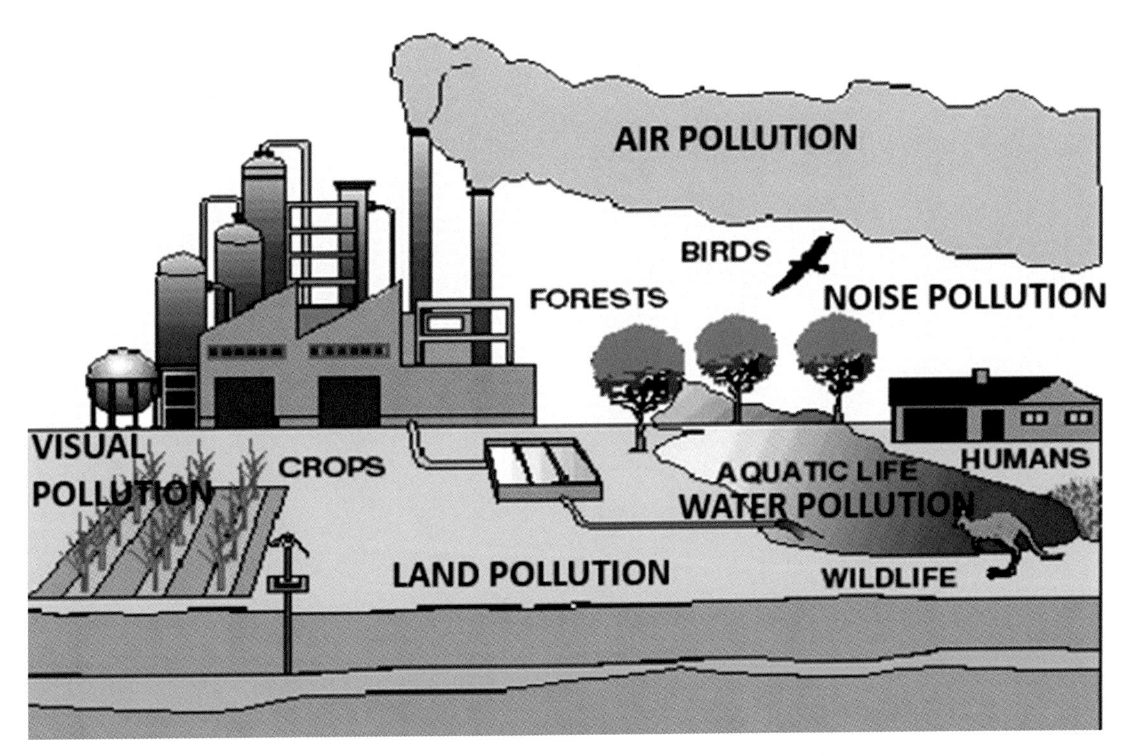

Figure 1.6 Different types of pollution [17].

- *Pollution:* This is basically undesirable change in physical, chemical, or biological characteristics of air, water, soil, or food that can adversely affect humans or other living organisms. Pollutants are mainly by-products of man's action and may include air pollution, water pollution, soil pollution, marine pollution, noise pollution, thermal pollution, industrial pollutants, agriculture pollutants, photochemical pollutants, nuclear hazards, and electromagnetic radiations. Figure

1.6 illustrates different types of pollution [17]. Once pollutants have become dispersed in the air, water, soil, or food at harmful levels, it is usually too expensive to reduce them to acceptable levels. In most cases, pollution clean-up often removes a pollutant from one part of the environment only to another.

- *Solid Waste Management:* This refers to collecting, treating, and disposing of solid material that is discarded or is no longer useful. The task of solid waste management presents complex technical challenges. The major sources of solid waste are households, agricultural fields, industries and mining, hotels and catering, roads and railways, hospitals and educational institutions, cultural centers and places of recreation and tourism, etc. Plastic waste is also a solid waste. Solid wastes can be classified as municipal waste, hospital waste, and hazardous waste. The hazardous waste can cause danger to health and environment.
- *Sustainability:* Sustainability is often regarded as meeting the needs of the current generation without compromising the ability for future generations to meet those same needs. It is the condition in which human needs are met in such a way that a human population can survive indefinitely. To achieve sustainability, we must think about how to implement it in all facets of life: buildings, streets, parks, roads, sidewalks, etc. Cities are key to sustainable development and sustainable future. They are responsible for making policies that affect sustainability. A sustainable society manages its economy and population size without exceeding all or part of the planet's ability to replenish its resources, and sustain human. A sustainable city, also referred to as an eco-city, designed with consideration for the triple bottom line: social, economic, and environmental impact, which is illustrated in Figure 1.7 [18]. The need for sustainable development is crucial to the future of mankind.

Figure 1.7 The three main components of sustainability [18].

Other components of environmental science include geosciences, atmospheric sciences, environmental law, environmental accounting, environmental physics, environmental biology, environmental economics, environmental hydrology, environmental health, environmental law, environmental management, environmental toxicology, etc.

1.5 CHALLENGES

Environmental scientists face a number of challenges. Environmental processes are complicated. It is often tacitly assumed that science works for the good of society, and that society should, therefore, embrace science. However, when environmentalists view science, they are torn between suspicion and trust: suspicion because some environmental crises have been caused by science, trust because the proof that these crises are real is provided by science. Environmentalists often view science with suspicion even when they see its friendliness [19].

Environmental scientists have always had to answer questions about their methods, techniques, data, assumptions, analysis, results, and conclusions since it is the nature of science to exchange and question research results. In addition to delays or suppression of certain information, many environmental scientists are reluctant to engage in certain research, or speak out on certain issues of environmental science, for fear of retribution. Many environmental science issues are reducible to differences of opinion on the appropriate methodology, degree of uncertainty or allegations of risk that cannot be proved or disproved. Misconduct in scientific research has come to limelight. An allegation of research misconduct can seriously prevent an environmental scientist from performing and distributing research results [20].

Despite growing exploitation of smart environment, the problems of human interaction with smart environment remain unresolved. Another interesting challenge is to consider not only the ability of the smart environment to fit user preferences but using smart environment to change behaviors in the individual.

1.6 CONCLUSION

Environmental science is an interdisciplinary study of how the earth works and how we can deal with the environmental issues we face. It helps us understand our environment and teaches us to use natural resources more efficiently. There is an ever demanding need for experts in this field because the environment is responsible for making our world beautiful and habitable. For this reason, environmental science is now being taught at high schools and higher institutions of learning [21]. Courses on environmental science will empower the youths to take an active role in the world in which they live. Education and awareness of the field of environmental sciences is becoming global and dynamic. More information about environmental science can be found in books in [9,12, 22-46] and related journals:

- *Energy & Environmental Science*
- *Environment International*
- *Environmental Development*
- *Environmental Modelling & Software*
- *Environmental Pollution*
- *Environmental Research*
- *Environmental Research Letters*
- *Environmental Science & Policy*
- *Environmental Science & Technology Letters*
- *Environmental Science & Technology*
- *Environmental Science & Engineering Magazine*
- *Journal of Environmental Science, Computer Science and Engineering & Technology*
- *Journal of Environmental Science: Current Research*
- *Current Opinion in Environmental Science & Health*
- *American Journal of Environmental Science and Engineering*
- *Global Journal of Environmental Science and Management*
- *International Journal of Environmental Research*
- *Journal of Ambient Intelligence and Smart Environments.*

REFERENCES

[1] "Environmental studies - Quick guide" https://www.tutorialspoint.com/environmental_studies/environmental_studies_quick_guide.htm

[2] M. N. O. Sadiku, U. C. Chukwu, A. Ajayi-Majebi, and S. M. Musa, "Environmental science: An introduction," *Journal of Scientific and Engineering Research*, vol. 7, no. 11, 2020, pp. 96-101.

[3] "Definitions of environmental health,"

https://www.neha.org/about-neha/definitions-environmental-health

[4] "Environment," Unknown Source

[5] "Environmental studies – Environment,"

https://www.tutorialspoint.com/environmental_studies/environmental_studies_environment.htm

[6] E. Ahmed et al., "Internet of things based smart environments: State-of-the-art, taxonomy, and open research challenges," *IEEE Wireless Communications,* vol. 23, no. 5, October 2016, pp. 101-6.

[7] M. N. O. Sadiku, S. M. Musa, and A. Ajayi-Majebi, **"**Smart environment: A primer," *International Journal of Environmental and Ecology Research*, vol. 1, no. 1, January 2019, pp. 25-26.

[8] M. N. O. Sadiku, O. D. Olaleye, and S. M. Musa, "Green environment," *International Journals of Advanced Research in Computer Science and Software Engineering,* vol. 9, no. 6, June 2019, pp. 51-54.

[9] G. T. Miller, Jr. and S. E. Spoolman, *Living in the Environment Concepts, Connections, and Solutions.* Belmont, CA: Brooks/Cole, Cengage Learning, 16th edition, 2009.

[10] "Environmental sciences,"

https://science.nd.edu/undergraduate/majors/environmental-sciences/

[11] "Environmental science," *Wikipedia,* the free encyclopedia,

https://en.wikipedia.org/wiki/Environmental_science

[12] Y. K. Singh, *Environmental Science.* New Delhi, India: New Age International, 2006.

[13] M. N. O. Sadiku, T. J. Ashaolu, A. Ajayi-Majebi, and S. M. Musa, "Environmental studies: An introduction," *International Journal of Scientific Advances,* vol. 1, no.3, Nov.- Dec., 2020, pp. 148-151.

[14] M. N. O. Sadiku, O. D. Olaleye, and S. M. Musa, "Environmental engineering: A primer*," International Journal of Trend in Research and Development,* vol. 6, no. 3, May- Jun. 2019, pp. 102-104.

[15] M. N. O. Sadiku, O. D. Olaleye, and S. M. Musa, "Environmental economics: A primer," *International Journals of Advanced Research in Computer Science and Software Engineering*, vol. 9, no. 7, July 2019, pp. 36-39.

[16] M. N. O. Sadiku, O. D. Olaleye, and S. M. Musa, "Environmental management: A primer," *International Journal of Trend in Research and Development,* vol. 6, no. 3, May- Jun. 2019, pp. 105-107.

[17] "Environmental pollutions," January 2019,

https://rainashafaa.blogspot.com/2019/01/environmental-pollutions.html?utm_source=feedburner&utm_medium=feed&utm_campaign=Feed:+blogspot/bRHdl/rainasblogs+(Welcome+to+the+Raina%27s+World)&m=1

[18] "The sustainable vernon initiative - How does it benefit you?"

http://www.tankerhoosen.info/news/upcoming_sustainable_vernon_2019.htm

[19] J. E. Foss, *Beyond Environmentalism: A Philosophy of Nature.* Hoboken, NJ: John Wiley & Sons, chapter 5, 2009.

[20] R. R. Kuehn, "Suppression of Environmental Science," *American Journal of Law & Medicine*, vol. 30, 2004, pp. 333-369.

[21] "Why study environmental science?" March 2019,

https://unity.edu/environmental-careers/why-study-environmental-science/

[22] P. Knoepfel, *Environmental Policy Analyses: Learning from the Past for the Future - 25 Years of Research.* Springer 2007.

[23] R. S. Khoiyangbam and N. Gupta, *Introduction to Environmental Sciences.* New Delhi, India: The Energy and Resources Institute, 2012.

[24] L. D. Williams, *Environmental Science Demystified.* New York: McGraw-Hill, 2005.

[25] C. Zehnder et al., *Introduction to Environmental Science.* University System of Georgia, 2nd Edition, 2018.

[26] A. M. Spooner, *Environmental Science for Dummies.* Hoboken, NJ: John Wiley & Sons, 2012.

[27] M. Allaby, *Basics of Environmental Science.* Routledge, 2nd edition, 2000.

[28] W. Cunningham and M. A. Cunningham, *Principles of Environmental Science.*

McGraw-Hill Education, 9th edition, 2019.

[29] D. E. Alexander and R. W. Fairbridge (eds.), *Encyclopedia of Environmental Science.* Netherlands: Kluwer Academic Publishers, 1999.

[30] C. M. Galanakis, *Innovation Strategies in Environmental Science. Elsevier, 2019.*

[31] B. J. Nebel and R. T. Wright, *Environmental Science: The Way the World Works.* Englewood Cliff, NJ: Prentice Hall, 4th edition, 1993.

[32] T. O'Riordan (ed.), *Environmental Science for Environmental Management.* Taylor & Francis, 2nd edition, 2014.

[33] W. P. Cunningham, M. A. Cunningham, and B. W. Saigo, *Environmental Science: A Global Concern.* McGraw-Hill Education, 15th edition, 2020.

[34] J. A. Tickner (ed.), *Precaution, Environmental Science and Preventive Public Policy.* Washington: Island Press, 2002.

[35] D. B. Botkin and E. A. Keller, *Environmental Science: Earth as a Living Planet.* John Wiley & Sons, 2002.

[36] R. McMullan, *Environmental Science in Building.* Palgrave Macmillan, 7th edition, 2017.

[37] R. T. Wright and D. Boorse, *Environmental Science: Toward a Sustainable Future.* Pearson; 13th edition, 2016.

[38] B. F. J. Manly, *Statistics for Environmental Science and Management.* Boca Raton, FL: CRC Press, 2008.

[39] E. D. Enger, B. F. Smith, and A. T. Bockarie, *Environmental Science: A Study of Interrelationships.* McGraw-Hill Education, 15th edition, 2018.

[40] J. R. Pfafflin and E. N. Ziegler (eds.), *Encyclopedia of Environmental Science and Engineering.* Boca Raton, FL: CRC Press, 5th edition, 2006

[41] P. J. Bowler, *The Norton History of the Environmental Sciences.* W. W. Norton & Co. Inc; 1993.

[42] G. T. Miller and S. Spoolman, *Environmental Science.* Cengage Learning; 16th edition, 2018.

[44] T. Forsyth, *Critical Political Ecology: The Politics of Environmental Science.* Taylor and Francis, 2003.

[45] M. L. McKinney et al., *Environmental Science: Systems and Solutions.* Sudbury, MA: Jones and Barlett Publishers, 4th edition, 2007

[46] M. D. Schwartz, *Phenology: An Integrative Environmental Science.* Springer, 2003.

CHAPTER 2

ENVIRONMENTAL NATURAL SCIENCES

"You cannot get through a single day without having an impact on the world around you. What you do makes a difference and you have to decide what kind of a difference you want to make." —Jane Goodall

2.1 INTRODUCTION

In a modern society, it is easy to forget that our society depends largely on the environmental processes that govern our world. The basis of our economy depends on the soils that sustain our agriculture, the rivers that provide our water, the minerals that provide the raw materials for the goods we consume, and the plants and animals that serve as our food [1]. Human inventiveness has introduced chemicals and materials into the environment. The natural environment is different from the built environment, which comprises the areas that are influenced by humans.

The natural sciences seek to understand how the universe around us works. There are five major branches: biology, chemistry, physics, astronomy, and earth science. Today, natural sciences are often divided into two main branches: life science and physical science. Life science is alternatively known as biology (botany and zoology), and physical science consists of physics, chemistry, earth science, and astronomy [2]. Physical science studies non-living systems, in contrast to the biological sciences.

Understanding environmental natural science is essential to stewardship of the environment and for addressing environmental problems like urban air pollution and for dealing with natural hazards such as floods and hurricanes. Almost all areas of natural science relate to environment and may have direct impact. The principles of natural science can help to evaluate and analyze environmental issues. Several volumes would be needed to cover all relevant principles of environmental natural science.

This chapter provides the fundamentals of environmental natural science and the effects of mankind's activities on the earth's systems. It begins by discussing the three major components of environmental natural sciences: environmental biology, environmental chemistry, and environmental physics. It covers how environmental natural science is practiced in nations around the world. The last section concludes with comments.

2.2 ENVIRONMENTAL BIOLOGY

Environmental biology studies the ways organisms, species, and communities influence, and are impacted by natural and human-altered ecosystems. It explores the interconnections among biology, ecology, evolution, environmental science, and conservation. It is a thriving field that is in dear need of enthusiastic, passionate, and well-trained professionals. Environmental biologists focus on the biology of ecosystems and environmental processes, causes and consequences of environmental change, and how environmental change impacts life on earth.

Environmental biology is a discipline in science at the intersection of environmental science, ecology, evolution, conservation, and global change. It examines the ways organisms, species, and communities influence, and are impacted by, natural and human-altered ecosystems. It covers all the fundamental concepts of the life sciences, including genetics, speciation, evolution, growth and differentiation, metabolism and bio-energetics, ecology, and behavior. It addresses many relevant issues that affect us on a daily basis such as energy conservation, air pollution, sustainable development, environmental toxins, etc. Figure 2.1 shows some aspect of environmental biology [3].

Figure 2.1 Conservation of natural habitats and species [3].

"Environmental biology" is often considered synonymous for "ecology." These two terms depict the same thing and they appear like the two sides of the same coin. Although both fields study ecosystems, environmental biology focusses more on the "biological organisms of environment." Ecology is literally the study of "houses" or more broadly, "environments." It is the study of how living organisms interact with their environment. Its main focus has been on various types of ecosystems – terrestrial, fresh water, marine and on how human activity has influenced these ecosystems. There are several types of ecology, such as ecosystem ecology, restoration ecology, physiological ecology, landscape ecology, animal ecology, practical ecology, and plant ecology. Ecologists essentially seek to explain interactions, interrelationships, behaviors, and adaptations of organisms. Figure 2.2 shows what ecology is all about [4].

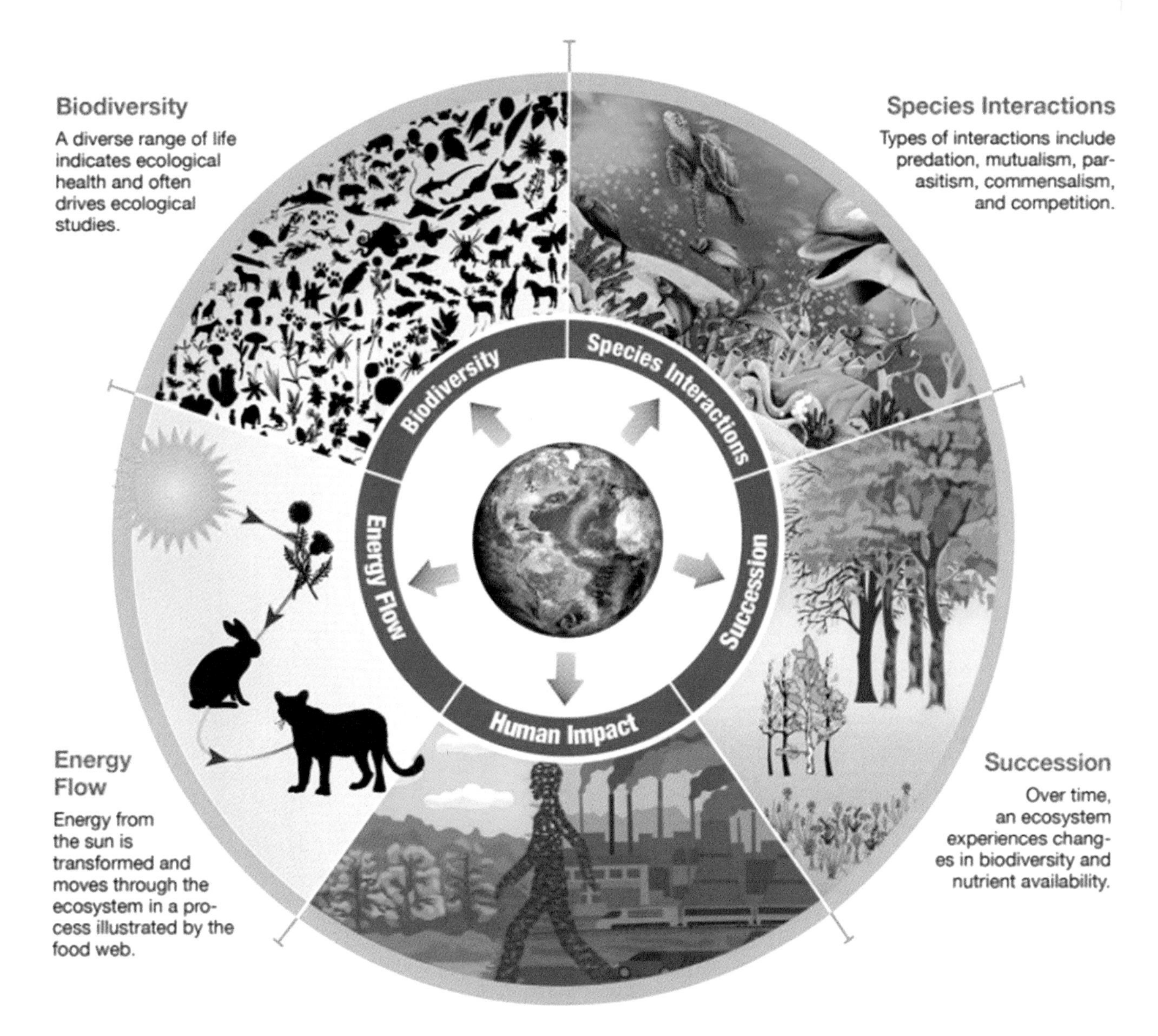

Figure 2.2 Ecology: The study of the place we live [4].

An environmental biologist studies the biology of specific organisms and their interactions with the environment. Environmental biologists are mainly responsible for monitoring environmental conditions and collecting water and soil samples from the field. Some monitor pollution levels to ensure compliance with state and federal laws. They assist companies to comply with regulations and conduct environmental impact assessments for their development projects. Environmental biologists often need to document their findings [5]. Becoming an environmental biologist has never been more exciting. Figure 2.3 shows a typical environmental biologist at work [6].

Figure 2.3 An environmental biologist at work [6].

Environmental biologists are needed in different areas such as food and agriculture, environmental education, natural resources sector, environmental consulting, non-governmental environmental organizations, environmental R&D, conservation and environmental protection, government agencies, biotechnology, and renewable energy. They work for their local environmental agencies, municipalities, the National Park Service, Forestry Service, Department of Commerce, the Environmental Protection Agency, the National Oceanic and Atmospheric Administration, National Institutes of Health, and other agencies. They also serve as field technicians, laboratory technicians, international relations, researchers for private or government laboratories, etc. [7].

A training on environmental biology could focuses on: (i) the biology of ecosystems and environmental processes, (ii) the causes and consequences of environmental change, and (iii) how environmental change impacts life on earth. It will provide the following skills [8]:

- Contemporary field and lab methods in ecology, evolution, and conservation biology
- Contemporary field and lab methods in environmental science for soil, water, air, and climate assessments
- Assessment of how toxic compounds impact life on earth, from individuals to species to communities
- Climate change impact assessments
- Climate change modelling with a focus on biodiversity impacts
- Environmental impact assessments and audits
- Design of applied environmental and ecological experiments
- Quantitative methods for collecting and interpreting ecological and environmental data
- Application of ecological research for environmental policy and decision-making
- Earth imaging, including Geographic Information Systems (GIS) and Remote Sensing (RS), for environmental problem solving and conservation
- Development of strong communication skills, including critical reading and writing

2.3 ENVIRONMENTAL CHEMISTRY

Environmental chemistry is the branch of chemistry that deals with the production, transport, and effects of chemicals in the water, air, earth, and natural environments. It is the field that studies chemical phenomena that occur in natural places. It involves understanding the chemical processes that occur in water, air, terrestrial, and living environments. Although some key chemical concepts are fundamental to understanding environmental chemistry, it is an interdisciplinary discipline that covers analytical chemistry, aquatic and soil chemistry, atmospheric chemistry, astrochemistry, engineering, mathematics, biology, geology, ecology, toxicology, and environmental sciences. Due to its multidisciplinary nature, environmental chemistry requires the contribution and collaboration of a wide variety of scientists to succeed. Figure 2.4 is a typical illustration of environmental chemistry [9].

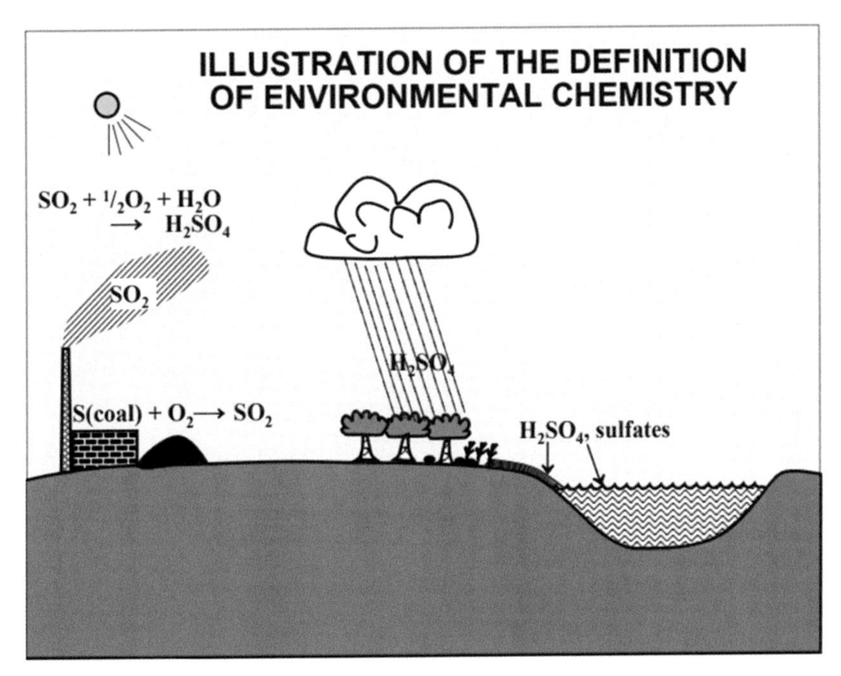

Figure 2.4 A typical Illustration of environmental chemistry [9].

Assessing the environmental impact may require a variety of experts: soil scientists to understand its fate in the soil, biologists to analyze the effect of the drug on living creatures, and atmospheric scientists to predict its transformations in the water and air [10].

Environmental chemistry should not be confused with green chemistry, which seeks to reduce potential pollution at its source, while environmental chemistry is the scientific study of the chemical phenomena that occur in natural places. Green chemistry is not confined to industrial sector. It applies to all areas of chemistry including organic chemistry, inorganic chemistry, biochemistry, analytical chemistry, and physical chemistry [11].

As a distinct field in chemistry, environmental chemistry emerged as a discipline when scientists started studying the occurrence of chemical in the natural environment. During the1960s, industries, agriculture, and households started to use more chemical compounds such as pesticides, detergents, polyester, synthetic rubber, etc. A major impetus for environmental chemistry dates from the discovery in the1970s of human health hazards caused by environmental pollution. Environmental chemists also began studying the effects of human-caused chlorofluorocarbons (CFCs) on the stratospheric ozone layer. Since then, environmental chemistry expanded to include the study of chemical compounds in water, soil, biological systems, etc. A contaminant or pollutant is a substance present in nature at a level higher than fixed levels or that would not otherwise be there.

There are four million known chemicals in the world today and another 30,000 new compounds are added to the list every year. When a chemical is released into the environment, it is distributed among the four major environmental compartments: (1) air, (2) water, (3) soil, and (4) living organisms. The environmental interest in acid-base chemistry is focused on the capacity of natural waters and soils to resist pH changes resulting from human activity [10].

Where do these chemicals come from? Some are from food additives, contamination, pollution, atmosphere, radioactive chemicals, etc. Some of these sources are discussed as follows [12,13].

- *Contamination:* This refers to the presence of one or more chemicals in the environment in higher concentrations than normally occurs but not high enough to cause harm and damage. Some industries emit metals that pollute the environment. For instance, lakes near Sudbury, Ontario, have been polluted by sulfuric acid, copper, nickel, and other metals which cause toxicity to plants and animals. Mercury contamination of fish is also a major challenge in many aquatic environments.
- *Pollution:* This is the effect of undesirable changes in our surroundings that have harmful effects on humans, plants, and animals. Pollution can poison soils and waterways, or kill plants, animals, and even humans. For example, long-term exposure to air pollution can lead to chronic respiratory disease, lung cancer and other diseases in humans. Air pollution can lead to reductions in agricultural yields, reduced growth and survivability of tree seedlings, and increased plant susceptibility to disease, pests, and harsh weather). There are several types of pollution: water pollution, air pollution, oil pollution, chemical pollution, etc. Sources of pollution include industrial waste disposal, water treatment plant, municipal sewage leakage, and sanitary landfills. These pollutants have damaged tens of thousands of lake and running-water ecosystems. A clear understanding of pollutants and their chemistry is essential for interpreting health effects, regulating emissions, and developing pollution-reducing technologies. Figure 2.5 shows that photochemical smog occurs where sunlight acts on vehicle pollutants [14].

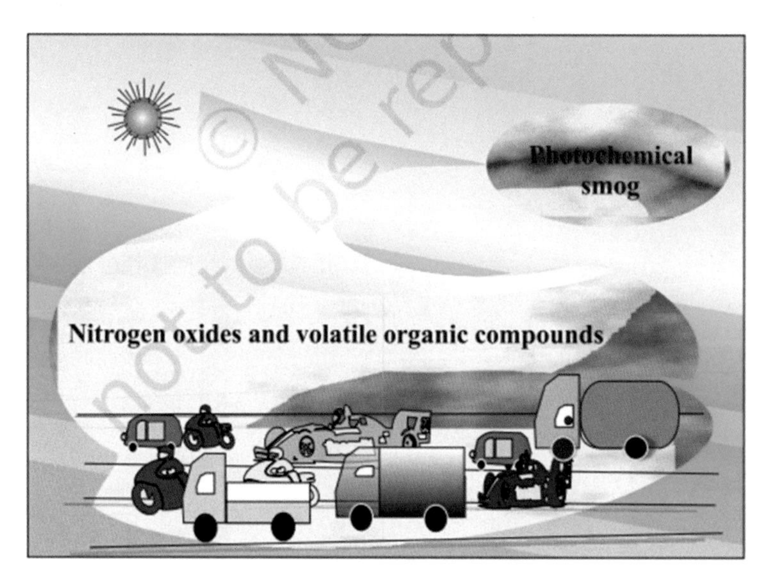

Figure 2.5 Photochemical smog occurs where sunlight acts on vehicle pollutants [14].

- *Atmosphere:* The atmosphere is a blanket of gases surrounding and protecting the earth. It can contain high concentrations of gases, vapors, or particulates that are potentially harmful to people or animals. The major gases in the atmosphere are nitrogen and oxygen, while the minor gases are argon, carbon dioxide, and some trace gases. Carbon dioxide and its carbonate minerals play an important role in environmental chemistry. Carbon monoxide is one of the most serious air pollutants.

Professional environmental chemists usually have at least a bachelor's degree. They are often employed by remediation firms, environmental consulting companies, state and federal regulatory agencies (such as Environmental Protection Agency (EPA)), manufacturing companies research centers, and academic institutions. They work in collaboration with biologists, geologists, atmospheric scientists, engineers, lawyers, and legislators. Figure 2.6 shows some environmental chemists in the field [15].

Figure 2.6 Some environmental chemists in the field [15].

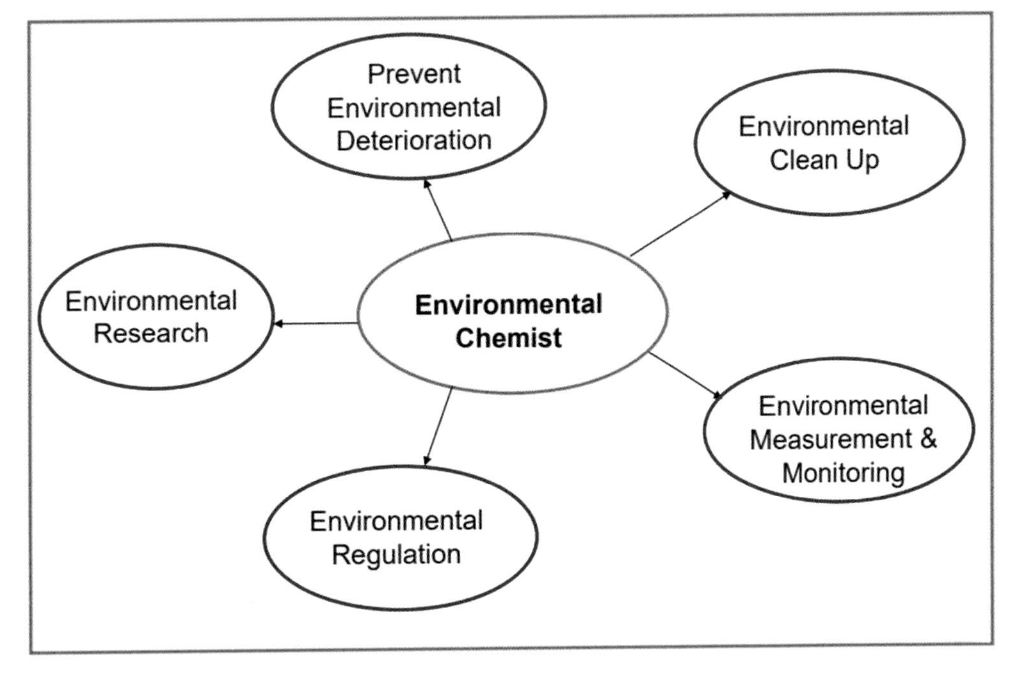

Figure 2.7 The mission of environmental chemists [16].

Figure 2.7 shows the mission of environmental chemists [16]. While responsibilities may vary significantly from one workplace to another, environmental chemists are responsible for the following [17]:

- Develop data collection methods and systems according to the elements that are under study
- Collect information from observations, samples, and specimens
- Record and manage records of observations, samples, and specimens in the lab and via fieldwork
- Use GIS and computer modeling to help forecast and analyze chemical impact
- Analyze literature, data, laboratory samples, and other sources of information to uncover primary, secondary, and tertiary chemical impacts
- Prepare reports and present research findings to internal and external stakeholders
- Communicate with team lead and executive through regular, scheduled reports and presentation of research findings
- Advise organizations and policymakers on the short and long-term impact and safety of chemicals in the environment
- Review research and literature in the field to stay abreast of current discoveries
- Classify contaminated soils as hazardous waste and manage their disposal
- Analyze new chemicals and their impact on the environment
- Conduct laboratory work, take measurements, interpret data, and use computers for environmental modeling
- Develop strong communication skills, including critical reading and writing

2.4 ENVIRONMENTAL PHYSICS

Physics is the study of the forces and laws of nature. Environmental physics is the branch of physics concerned with the measurement and analysis of interactions between organisms and their environment. It is often regarded as the fundamental science, because all other natural sciences (such as astrophysics, geophysics, chemical physics and biophysics) apply the principles and laws of physics. Physics relies heavily on mathematics as the tool for problem formulation and quantification of principles. Physical principles are largely derived from direct observation and experimentation of nature. Physics is a broad field including mechanics, optics, electricity, magnetism, electromagnetics, thermodynamics, and quantum mechanics.

A physicist is someone who studies the forces, laws, and behavior of nature to understand how things work. Most physicists use math to describe theories and processes, use complex calculations, use software packages to simulate physical systems, observe and measure physical phenomena in laboratories, conduct research and development, and share their research results with others at conferences. Other physicists with advanced degrees teach and do research as faculty members at colleges and universities.

Environment may be regarded as the medium in which any entity finds itself. Environmental physics is essentially a physical science. It applies the theoretical and experimental techniques of optical, surface, and condensed matter physics to the environmental problems facing us today. It embraces the following concepts: human environment and survival physics, built environment, urban environment, renewable energy, remote sensing, weather, climate change, environmental health and environmental control [18]. It deals with atmospheric science, pollution detection and remediation, dynamical processes at land-water, colloidal science, and biological effects of pollution, and bioremediation science, weather forecasting, and environmental control. To fully understand the complexities of the environment and to address environmental problems effectively, the underlying physics must be combined with biology, chemistry, and geology. Effective management of human interaction with an environmental system requires simultaneous progress on several fronts [19]. Figure 2.8 illustrates the field of environmental physics [20].

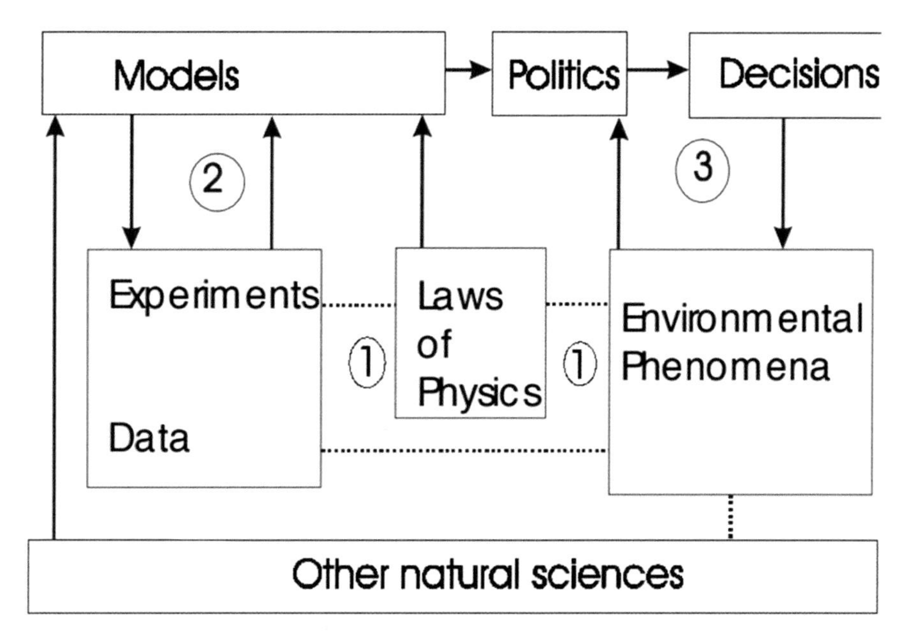

Figure 2.8 The field of environmental physics [20].

Environmental physicists apply physical principles to environmental problems. They seek to understand the mechanisms by which environmental change impacts humans and other organisms. They need a sound understanding of basic physics, properties of matter, thermodynamics, and fluid dynamics. They may have careers in climate modeling, pollution control, energy conservation, renewable energy, and waste disposal. Aspiring environmental physicists may also consider specializing in condensed matter physics, electronics, ecology, environmental assessment, and environmental biology. Environmental jobs are increasing in demand [21]. Figure 2.9 shows some environmental physicists at work [22]. The skills expected of environmental physics graduates include [23]:

Figure 2.9 Some environmental physicists at work [22].

- Apply physical principles to problems and formulate solutions
- Integrating theoretical approaches
- Advanced mathematical ability
- Team-based research and ability to work in multi-disciplinary groups
- Design and execute experiments
- Write technical reports and project proposals relevant to given audience
- Demonstrate ethical scientific behavior
- Utilize qualitative and quantitative analysis and problem solving

- Geographic Information Systems (GIS) and Remote Sensing (RS) for solving problems in the environmental sciences
- Qualitative and quantitative approaches for environmental impact assessments & audits
- Conduct experiments, take measurements, interpret data, and use computers for environmental modeling.

2.5 GLOBAL ENVIRONMENTAL NATURAL SCIENCES

Scientists as well as politicians globally are looking for how to solve environmental problems. The UN's Sustainable Development Goals include universal calls to action to protect life on land and in water. Environmental chemistry is used by the United States Environmental Protection Agency, the Environment Agency in England, Natural Resources Wales, and other national agencies. It is true that industrialized nations have population growth rates lower than those in developing nations. Yet the average person in the industrialized world consumes about 60 times more resources than someone in the developing world.

Collaboration between natural and social scientists is desirable since it exposes students to different approaches to environmental issues. Environmental biologists, environmental chemists, and environmental physicists worldwide can play an important role in helping ensure nations maintain high-quality supplies of water in the nation's rivers, lakes, and underground reservoirs.

The field of environmental natural science is broad and global. Although environmental impact varies significantly between nations, global inequalities in resource consumption and purchasing power mark the clearest dividing line between the haves and the have-nots. Natural resources of planet earth are under severe threat by global warming. Education and training on environmental natural science will vary depending on the location and expertise of the faculty.

. We now consider how environmental natural science is handled in some countries.

- *United States:* United States and other developed nations tend to have the most problem with the degradation of ecosystems and harming the world's poorest people. Environmental justice is a social movement that attempts to identify and remedy environmental injustices: when environmental benefits and burdens are unfairly distributed. The country's aging infrastructure will also need to be replaced. Developed nations, like US, do not want to be too much dependent on the willingness of other nations to supply them with oil and gas. In 1988, the United States ratified the Montreal Protocol, an international agreement ratified by 196 states to phase out the industrial production of chlorofluorocarbons (CFCs). According to some data on chemical exposure collected by the U.S. Centers for Disease Control and Prevention, the average US citizen has detectable levels of more than 100 xenobiotic compounds in his/her blood. Some retailers such as Wal-Mart and Staples use some assessments to certify the superior environmental performance of the products they sell [24].
- *United Kingdom:* The world's population has changed over time and the population of the UK is no exception. The carrying capacity of the UK for people is difficult to determine and depends on whether one means the carrying capacity with respect to sustainable food production or renewable energy resources. Like UK, governments all over the world are stimulating renewable energies. Some have argued that the quality of life would be much better in the UK if there were only half or a third the number of people there are today. In that case, there would be less pollution, more room for wildlife, and less traffic jams [25]. Accurate measurements of pollutants are vital for ensuring compliance with national and international air quality directives. The United

Kingdom has around 300 air quality monitoring sites measuring a variety of pollutants, including ozone, nitrogen oxides, sulfur dioxide, carbon monoxide, and particulates. The UK air quality target for nitrogen dioxide is an annual mean of 40 micrograms per cubic meter. The UK's professional and learned societies (including the sciences, social sciences, arts, humanities, medicine, and engineering) have endorsed a communiqué on climate change calling for government action [26].

- *Australia:* The average yield of Australia's major grain (wheat) rose at its fastest rate ever during the last decade. The environmental biology behind this advancement was predominantly ecological and nutritional. Applying nitrogen fertilizers became less risky. This allowed farmers to use much more fertilizers and thereby producing more. In spite of Australia's reputation for being drought prone, its crop yields have not hitherto been typically limited by water — poor health and poor nutrition have been more influential [27].
- *China:* In Asia, environmental chemists have been able to document pollution from electronic waste recycling in Southern China. Recent research indicates that the levels of PfAAs (a class of chemicals with unique water-, dirt-, and oil-repelling properties) in Chinese infants are higher than those for infants from other countries, suggesting that the use of perfluorinated (PfC)-containing products may be increasing in China [28].
- *India:* Incidences of environmental issues in India include white marble of Taj Mahal and leakage of methyl isocyanate (MIC) vapors at Bhopal in 1984. Pollution of river water in India, use of plutonium or other isotopic fuel-based breeder/nuclear reactors for energy production, use of dangerous artificial food additives, and ozone hole in the Antarctic and Arctic regions are some typical chemical issues that need to be resolved critically. Pollutants such as sulfur dioxide and nitrogen oxides can form acid rain, which pollutes soil and water and damages buildings such as the Acropolis and the Taj Mahal [13].

2.8 CONCLUSION

Environmental issues are now part of every career path and employment area. These issues are immensely complex, involving aspects of history, philosophy, behavior, science, economics, social justice, and politics. To understand and explore relationships between organisms and their environment, environmental scientists should be familiar with the main concepts of the environmental natural sciences. Environmental biology is the branch of biology which focuses on how organisms interact with the environment, and how they adapt to changing environments. Environmental chemistry focuses on how chemicals are formed, how they are introduced into the environment, and the effects they have on the environment and mankind. Environmental physics is the measurement and analysis of interactions between organisms and their environments.

The job outlook for environmental natural scientists is growing faster than the average for all occupations. To help meet the demand for specialized expertise in environmental natural science, several universities in the US are offering courses and degrees in the area.

The National Association of Environmental Professionals (NAEP) is a multidisciplinary association for all types of environmental professionals. The Society of Environmental Toxicology and Chemistry (SETAC) is a global professional organization that provides a forum for sharing ideas and promotes multidisciplinary approaches to solving environmental problems. More information about environmental natural sciences can be found in the books in [25,28-78] and the following related journals:

- *Journal of Environmental Biology*
- *Expert Opinion on Environmental Biology*

- *The Open Journal of Environmental Biology*
- *Advances in Environmental Biology*
- *Environmental Chemistry*
- *Environmental Chemistry Letters*
- *Environmental Briefs*
- *Frontiers in Environmental Chemistry*
- *Journal of Environmental Chemistry*
- *Journal of Environmental Chemical Engineering*
- *Journal of Environmental Chemistry and Ecotoxicology*
- *Journal of Environmental Chemistry and Toxicology*
- *International Journal of Environmental Chemistry*
- *International Journal of Environmental Analytical Chemistry*

REFERENCES

[1] "Course Syllabus: BIOL 2650: Environmental Biology,"

https://www.csustan.edu/sites/default/files/groups/Department%20of%20Biological%20Sciences/documents/syllabi/spring_2013/biol/biol2650.001.pdf

[2] "Natural science," *Wikipedia*, the free encyclopedia

https://en.wikipedia.org/wiki/Natural_science

[3] NSF, "Long term research in environmental biology,"

https://www.nsf.gov/news/newsmedia/ENV-discoveries/LTREB-discovery-series.jsp

[4] https://www.pinterest.com/pin/200199145914562663/

[5] M. N. O. Sadiku, T. J. Ashaolu, A. Ajayi-Majebi, and S. M. Musa, "The essence of environmental biology," *International Journal of Scientific Advances*, vol. 1, no.3, Nov.- Dec., 2020, pp. 144-147.

[6] "A career in the wild world of biology,"

https://www.aibs.org/careers/?gclid=EAIaIQobChMI8casivDw7AIVw5JbCh0eYghNEAAYASAAEgLDCPD_BwE

[7] "Environmental biology degree,"

https://www.environmentalscience.org/degree/environmental-biology

[8] "What is an environmental biologist?"

https://www.environmentalscience.org/career/environmental-biologist#:~:text=Environmental%20biologists%20are%20mainly%20responsible,soil%20samples%20from%20the%20field.

[9] "Narrative for a lecture on environmental chemistry,"

https://www.asdlib.org/onlineArticles/ecourseware/Manahan/EnvChBasicsCondensedLect.pdf

[10] "Environmental chemistry," *Wikipedia*, the free encyclopedia,

https://en.wikipedia.org/wiki/Environmental_chemistry

[11] M. N. O. Sadiku, S. M. Musa, and O. M. Musa, "Green chemistry: A primer," *Invention Journal of Research Technology in Engineering and Management*, vol. 2, no. 9, September 2018, pp. 60-63.

[12] "Environmental chemistry," Unknown source.

[13] N. Gupta, R.S. Khoiyangbam, and N. Jain, "Environmental Chemistry," Chapter · January 2015,

file:///C:/Users/12818/Downloads/Ch_2-EnvChemistry%20(1).pdf

[14] "Environmental chemistry,"

https://ncert.nic.in/textbook/pdf/kech207.pdf

[15] "Environmental chemistry major,"

https://www.hartwick.edu/academics/majors-and-minors/environmental-chemistry-major/

[16] H. Al-Najar, "Chapter 1: Introduction to environmental chemistry,"

http://site.iugaza.edu.ps/halnajar/files/2014/09/Chapter-1.-Introduction-to-Environmental-Chemistry.pdf

[17] "What is an environmental chemist?"

https://www.environmentalscience.org/career/environmental-chemist

[18] "Form six environmental physics,"

https://msomiexpress.wordpress.com/2019/08/03/form-six-environmental-physics/

[19] "The environment, " *Physics in a New Era: An Overview.* Washington, DC: The National Academies Press, chapter 7, 2001.

[20] E. Boeker, R. V. Grondelle, and P. Blankert, "Environmental physics as a teaching concept," *European Journal of Physics,* July 2003.

[21] "What is a physicist?" Unknown Source

[22] "Environmental physics,"

https://www.utsc.utoronto.ca/admissions/programs/environmental-physics

[23] "Career options after environmental physics,"

https://utsc.utoronto.ca/aacc/career-options-after-environmental-physics

[24] P.S. Verma and V. K. Agarwal, *Environmental Biology: Principles of Ecology.* New Delhi, India: Publisher S. Chand & Co, 2000.

[25] M. Reiss and J. Chapman, *Environmental Biology.* Cambridge University Press, 2nd edition, chapter 1, 2000.

[26] Royal Society of Chemistry, "Environment,"

https://www.rsc.org/campaigning-outreach/global-challenges/environment/

[27] J. B. Passioura, "Review: Environmental biology and crop improvement," *Functional Plant Biology,* vol. 29, no. 5, May 2002, pp. 537-546.

[28] K. Betts, *A Survey of Environmental Chemistry Around the World: Studies, Processes, Techniques, and Employment.* American Chemical Society, 2014.

[29] M. D. Goldfein and A. V. Ivanov, *Applied Natural Science: Environmental Issues and Global Perspectives.* Apple Academic Press, 2016.

[30] J. Maskall and A. Stokes, *Designing Effective Fieldwork for the Environmental and*

Natural Sciences. Plymouth: The Higher Education Academy Subject Centre for Geography, 2008.

[31] M. R. Fisher, *Environmental Biology.* Open Oregon Educational Resources, 2018.

[32] B. Bhatia, G. S. Chhina, and B. Singh, *Selected Topics in Environmental Biology.* Pergamon, January 1977.

[33] M. Gupta, *Fundamentals of Environmental Biology.* I K International Publishing House, 2018.

[34] D. A. Vaccari, P. F. Strom, and J. E. Alleman, *Environmental Biology for Engineers and Scientists.* Wiley-Interscience, 2005.

[35] F. Deeba, *Ecology and Environmental Biology.* Centrum Press, 2017.

[36] I. H. Zaheed, *A Text Book on Environmental Biology.* New Delhi, India: Discovery Publishing House, 2013.
[37] L. M. Lynn, *Environmental Biology and Ecology Laboratory Manual.* Kendall Hunt Publishing; 6th edition, 2016.

[38] H. R. Singh, *Environmental Biology.* S. Chand Publishing, 2nd edition, 2004.

[39] G. Tripathi, *Modern Trends in Environmental Biology.* CBS Publishers & Distributors, 2002.

[40] J. Ramsay and J. Schroer, *Environmental Biology.* Kendall Hunt Publishing Co., 2020.

[41] S. Z. Ali, *Environmental Biology.* Akhand Publishing House, 2019.

[42] P. S. Nobel, *Environmental Biology of Agaves and Cacti.* Cambridge University Press, 2003.

[43] A. J. Bailer, *Statistics for Environmental Biology and Toxicology.* Boca Raton, FL: CRC Press, 2020.

[44] P. D. Sharma, *Environmental Biology and Toxicology.* Rastogi Publication, 2005.

[45] K. Betts, *A Survey of Environmental Chemistry Around the World: Studies, Processes, Techniques, and Employment.* American Chemical Society, 2014.

[46] Royal Society of Chemistry, "Environment,"

https://www.rsc.org/campaigning-outreach/global-challenges/environment/

[47] J. G. Ibanez et al., *Environmental Chemistry Fundamentals.* Springer, 2007.

[48] J. G. Ibanez et al., *Environmental Chemistry: Microscale Laboratory Experiments.* Springer, 2008.

[49] S. E. Manahan, *Environmental Chemistry.* Boca Raton, FL: CRC Press, 10the edition, 2017.

[50] J. W. Moore and E. A. Moore, *Environmental Chemistry.* Elsevier, 1976.

[51] R. A. Hites, *Elements of Environmental Chemistry.* John Wiley & Sons, 2007.

[52] R. M. Harrison, *Understanding Our Environment: An Introduction to Environmental Chemistry and Pollution.* The Royal Society of Chemistry, 1999.

[53] J. Wright, *Environmental Chemistry.* Taylor & Francis, 2005.

[54] B. R. Sudani, *Analytical Environmental Chemistry.* New Delhi, India: AkiNik Publications, 2018.

[55] B. Pani, *Textbook of Environmental Chemistry.* I K International Publishing House, 2017.

[56]1. G. W. Vanloon and S. J. Duffer, *Environmental Chemistry - A Global Perspective.* Oxford University Press, 2000.

[57] V. K. Ahluwalia, *Advanced Environmental Chemistry.* The Energy and Resources Institute (TERI), 2017.

[58] E. R. Weiner, *Applications of Environmental Chemistry: A Practical Guide for Environmental Professionals.* Boca Raton, FL: Lewis Publishers, 2010.

[59] F. W. Fifield and W. P. J. Hairens, *Environmental Analytical Chemistry.* Black Well Science Ltd, 2nd edition, 2000.

[60] G. Hanrahan, *Key Concepts in Environmental Chemistry.* Waltham, MA: Academic Press, 2012.

[61] J. E. Andrews et al., *An Introduction to Environmental Chemistry.* Blackwell Science Ltd, 2nd edition, 2004.

[62] C. Baird, *Environmental Chemistry.* New York: Free-man and Company, 1995.

[63] Z. He (ed.), *Environmental Chemistry of Animal Manure.* Nova, 2011.

[64] E. Boeker and R. V, Grondelle, *Environmental Physics: Sustainable Energy and Climate Change.* John Wiley & Sons, 4th edition, 2013.

[65] A. W. Brinkman, *Physics of the Environment.* Imperial College Press, 2008.

[66] C. Smith, *Environmental Physics.* New York: Routledge, 2004.

[67] N. Mason and P. Hughes: *Introduction to Environmental Physics: Planet Earth, Life and Climate.* Taylor and Francis, 2001.

[68] D. Hillel, *Environmental Soil Physics.* San Diego, CA: Academic

Press, 2003.

[69] K. Forinash, *Foundations of Environmental Physics: Understanding Energy Use and Human Impacts.* Washington DC: Island Press, 2010.

[70] P. Singh and T. A. Wani, *Basic Environmental Physics.* Pragati Prakashan, 2016.

[71] *Encyclopaedia of Introduction to Environmental Physics: Planet Earth, Life and Climate* (4 Volumes). Hillingdon, UK: Publisher Koros Press Limited, 2015.

[72] R. E Robson and D. Blake, *Physical Principles of Meteorology and Environmental Physics: Global, Synoptic and Micro Scales.* World Scientific Publishing, 2008.

[73] F. Borghese, P. Denti, and R. Saija, *Scattering From Model Nonspherical Particles: Theory and Applications to Environmental Physics.* Springer, 2007.

[74] D. Hillel, *Environmental Soil Physics: Fundamentals, Applications, and Environmental Considerations.* Academic Press, 2004.

[75] J. L. Monteith and M. H. Unsworth, *Principles of Environmental Physics.* Elsevier, 4th edition, 2013.

[76] V. Faraoni, *Exercises in Environmental Physics.* Springer, 2006

[77 P. Hughes and N.J. Mason, *Introduction to Environmental Physics: Planet Earth, Life and Climate.* London, UK: CRC Press, 2014.

[78] Calvin W. Rose, *An Introduction to the Environmental Physics of Soil, Water and Watersheds.* Cambridge University Press, 2012.

CHAPTER 3

ENVIRONMENTAL SOCIAL SCIENCES

"The environment is where we all meet; where we all have a mutual interest; it is the one thing all of us share." -Lady Bird Johnson

3.1 INTRODUCTION

Humans have a huge impact on the earth and the environment. Environmental degradation has been attributed to overpopulation, overconsumption, pollution, deforestation, landfills, land disturbances, and other natural causes.

It is well known that industrialized nations consume the lion share of the earth's resources. Yet efforts to address consumption are also strongly opposed by politically powerful interests and short-sightedness. For example, Americans represent around 5 percent of the earth's population, and yet consume 25 percent of the resources [1].

Although environmentalists generally agree that reducing our impact on the human environment (i.e., physical, biological, social, and cultural) requires a reduction of population growth and levels of consumption, there is no consensus on how we can achieve this. Historically, biologists have not been trained in social sciences, and social scientists have not generally been interested in the environmental issues. This situation is improving, as the social sciences are becoming increasingly ecologically informed and new interdisciplinary fields are emerging [2]. Environmental social science has emerged as the interdisciplinary study of what governs people's behavior and beliefs and how they organize themselves in relation to the environment.

This chapter provides an introduction on social science approaches to environmental issues. It begins with the connection between social sciences and environment. It then presents various components of environmental social science. It covers the role of environmental social scientists. It discusses global environmental social sciences. It addresses the benefits and challenges of environmental social sciences. It concludes with comments.

3.2 SOCIAL SCIENCES AND ENVIROMENT

Social science is a wide area that covers many subjects that deal with humans. It includes geography, economics, sociology, law, anthropology, and political science.

It is increasingly enrolled in policy, market, and activist endeavors in the area of environmental issues. Some areas within social science of relevance to the environmental sciences include [3]:

- political economy, global-to-local political relationships, the development of power asymmetries – particularly in regard to access to resources, health, etc.
- equity, justice, property rights, and social movements

- trust, governance, conflict
- consumption, social, cultural and symbolic capital
- institutions
- migration, indigeneity, and ethnicity
- markets, commodities, motivations, values and cosmologies, and time horizons

Environmental social sciences (ESS) are concerned with how humans perceive, understand, and act or influence the environment. It addresses the complex relationship between people and the environment by exploring issues related to climate change, carbon emissions, global food and water distribution, natural resource management, and conservation decision-making. It is the interdisciplinary study of what governs people's behavior, beliefs, and ways of organizing themselves in relation to the natural environment. It increasingly includes areas such as environmental history, environment sociology, and environmental philosophy. It promotes the idea that the social, cultural, economic, political science, historic, institutional, psychological, sociological, human ecology, and environmental factors are all important for consideration. The factors are needed both to describe and understand human-environment interactions [4].

3.3 COMPONENTS OF ENVIRONMENTAL SOCIAL SCIENCES

Environmental social science is the application of social science and its application to our understanding of environmental issues. It explores complex social forces shaping how humans interact with the environment. It is the broad, transdisciplinary study that covers anthropology, economics, geography, history, political science, psychology, and sociology. Thus, the components of environmental social science include environmental economics, environmental sociology, environmental philosophy, environmental geography, environmental history, environmental policy, environmental movement, etc.

- *Environmental Economics:* This is a relatively new sub-discipline of economics that really took off in the 1960s. It adopted the main models and methodology of mainstream economics at that time. Until the early 1900s, most economists were social scientists who analyzed economic problems from a social science point of view. Modern environmental economics definitely entered the scene in the 1960s. Today, environmental economics is largely based on the heavily criticized standard neoclassical model. Environmental problems and their solutions are ultimately social and behavioral in nature. For this reason, behavioral environmental economics is currently gaining ground [5].
- *Environmental Sociology:* This is the study of the relationship between modern societies and the environment. It includes topics such as the environment, environmental social movements, inequality, social change, and justice. Sociology is known for its academic debates, which are vehicles to accumulate understanding and interpretation of a constantly changing modern order. Sociologists have been accused of sociological foot-dragging on environmental matters. This has spurred some sociological experts to actively become advocates, and even "hucksters," for the benefit of technological innovation and economic development [6].
- *Environmental Movement:* The period of environmental movement or activism debate that dates from around the end of the 1960s. Environmental movement or activism attempts to change how contemporary societies use scarce materials while recognizing that we are in the midst of an irreversible process of global transformation. Environmental social movements may influence the construction of 'nature' and the perceived threat to environment. Many

conventional approaches to environmental activism assume that scientific knowledge is non-negotiable and that environmentalism will lead to a greater democracy. There is hidden biases in environmentalism. The environmental norms developed in developed nations may not be applicable to developing nations [7].

- *Environmental Policy:* Researchers have analyzed the interactions between social science and environmental policy, which highlights the tensions across policy agendas such as environmental protection, economic growth, and trade liberalization. Some argue that environmental governance helps governments to deal with the growing number of conflicts and protests. Scholarship across a range of disciplines provides a wealth of insights for thinking critically about efforts to influence environmental politics and policy [8]. With the planet continually facing environmental threats, improving environmental policy making relies on the insights of social science research as well as those of the natural sciences.

3.4 ENVIRONMENTAL SOCIAL SCIENTISTS

Environmental social scientists work within the fields of anthropology, economics, geography, history, political science, psychology, and sociology. Each of these disciplines has unique perspectives and makes noteworthy contributions to environmental studies and also complement one another. Environmental social scientists focus on human – environmental relationships. They are effectively making positive changes in the environment. They deal with the idea of "environmental justice" which connects issues in the field of social justice with the environment. Figure 3.1 shows some environmental social scientists at work [9].

Figure 3.1 Some environmental social scientists at work [9].

Social scientists from the fields of anthropology, economics, geography, political science, psychology, and sociology need to collaborate more effectively with colleagues from the natural sciences. They should become engaged with research beneficial to the society first and then let the natural science do the complementing. The natural scientists and social scientists need should work on multi-disciplinary research projects together [10]. Figure 3.2 shows the interactions between natural and social sciences [11].

Social sciences and environmental science have a lot in common. Social sciences include economics, public policy, sociology, management, and psychology. They look at humans while environmental looks at the earth and living things. They seek solutions to problems like climate change, emerging pandemic disease, and large-scale extinction, global food and water distribution, natural resource management, etc. Ideas related to exploring human and animal interactions within the natural world, have become prominent in environmental ethics.

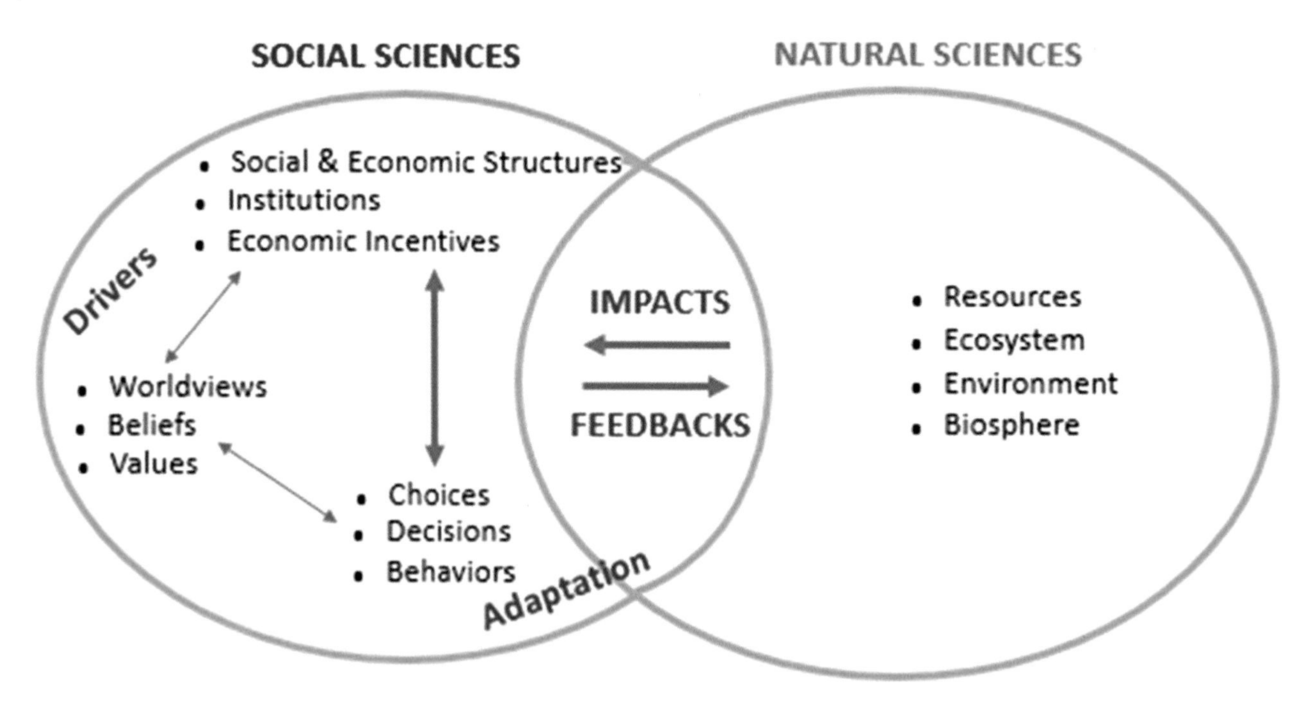

Figure 3.2 Interactions between natural and social sciences [11].

3.5 GLOBAL ENVIRONMENTAL SOCIAL SCIENCES

In September 2015, experts from eleven countries in Europe and America met in Brazil to discuss how the interactions between social science and humanities and environmental science researchers are opening new pathways to provide innovative solutions to global problems [12].

Some environmentalists have argued that environmental problems are increasingly global in nature. Environmental global changes have led to an increasingly interconnected world, increasing exchange of people and goods. They increase the capacity of a nation to provide people with goods and services. There is increasing evidence that the environmental norms developed in Europe and North America have produced environmental science in a way that may not be applicable to other societies worldwide. We consider how some nations handle environmental social sciences.

- *United States:* The National Environmental Policy Act (NEPA) was passed in 1969 in response to public concerns about the degradation of the quality of the human environment. NEPA establishes a national policy for the protection and maintenance of the environment by providing a process which all federal agencies must follow. Environmental Assessments (EAs) are used to determine if significant environmental impacts would occur when a project is funded by a federal agency. EAs are concise documents that include analysis regarding the significance of environmental impacts of the proposed project, a listing of alternatives, and a listing of agencies and persons consulted [13]. The United States has seen many environmental cases, where polluter and victim are part of the same community.

- *Thailand:* Before the 1980s, environmental concern in Thailand was restricted to the activities of urban groups and interest in deforestation or forest conservation. The environmental movement in Thailand has become a significant force in recent years. The movement has drawn in a wide range of social, economic, and political actors. Yet it has also maintained its role as a significant challenge to dominant patterns of development and vested interests embodied in the status quo. Although environmental social movements in Thailand may apparently be "of a class, not on behalf of a class," the discourses of environmentalism have been adopted and used by different actors to support political objectives [7].
- *Africa:* In Africa, research is still dominated by the natural sciences. African social scientists, in collaboration with other researchers, are beginning to focus on the implications of climate change for livelihoods and development. Some nations, including African nations, nowadays seek foreign aid, which is almost always conditional on the application of environmental measures and reduction or avoidance of environment problems.
- *Japan:* The Japanese have seriously questioned the national government's invasive reach into local communities in the name of national development. Japanese case of mercury poisoning in Minamata, produced serious illness within the local population. This well-known case represents the primitive logic of pollution, whereby the problem is visited on an area, region or community by industrial activities carried on within that area. The poisoned Minamata fish could affect rich and poor alike. Due to the vested interests of the Chisso Company and the Japanese government, it took decades before the company, government, and society to acknowledge the problem and found the company responsible for the mercury poisoning. Minamata case represents the more classic case of a pollution industry denying that the industry is causing harm [14].
- *India:* In Delhi, the capital and one of the famous cities of India, the working class has perceived the misplaced priorities of city officials and influential environmental advocates. There is widespread frustration among environmental advocates towards the Supreme Court decision to promote pollution control that benefits the wealthy and privileged few at the expense of the vulnerable working class. By forcing polluting companies to relocate, the environmental campaigners destroyed the livelihood for many in an attempt to clean the air [14].

3.6 BENEFITS

Research has shown that healthy natural environments provide physical and psychological benefits for those who spend time in them. Social scientists have been advising on environmental problems for decades. They are specialists who are seriously concerned about the fate of our planet, its natural resources, and its inhabitants. They carefully provide analysis, evidence, and advice on environmental problem from the social science viewpoint. Since human behavior is central to many aspects of most environmental social sciences, the contributions of social scientists to work on environmental problems is important. There are benefits for social scientists who collaborate with natural scientists in solving environmental problems.

There is an urgency to changing norms because of mounting environmental problems such as climate change, water security, depletion of natural resources, pollution, energy, land, etc. Understanding these environmental problems significantly help state actors (e.g., governments, regulatory bodies), businesses, and, the general public. This will help them adopt environmental management measures which will ensure that the polluter pays a fine to minimize unwanted impacts.

3.7 CHALLENGES

The interaction between the "environmental" and the "social" still remains an uncharted terrain. Natural science and social science currently lack the integration necessary to understand ourselves or our relationship with the rest of nature. Solving environmental problems is usually complex and require interdisciplinary research involving the social and environmental sciences. Reducing human impacts and developing more sustainable environmental practices is difficult to achieve without using critical social science perspectives. However, attempts have been made to analyze the environmental–social interface in practice. Although the stakes for involving the social sciences in environmental research are high, natural scientists continue to conduct environmental social science research without any experience in the social sciences. The social sciences and natural sciences must be closely integrated to tackle the transformation needed to effect global environmental change. Rather than working with social scientists, many natural scientists continue to step over disciplinary boundaries to conduct attitude studies [15].

A great majority of social scientists cannot compete with natural scientists with respect to research funding or publications. Another barrier to interdisciplinary engagement for social scientists is the tendency for scientists from high-prestige disciplines to dabble into social science. Although graduate students may choose a course of study that aligns with their passions, with so many paths available, it can be hard to know which one to choose. Developing nations lack funding for environmental social science research. There is virtually no funding for social science research on environmental issues.

3.8 CONCLUSION

Environmental problems are human problems. There is a demand for better integration of social sciences into conservation training and practice. People rely on a healthy natural environment for clean air and water, food, shelter, and materials. Local natural environments are a source of pride, and people are willing to spend their vacations traveling to view natural landscapes [16]. The environmental social sciences rely on cultural, economic, historic, political, psychological, and sociological factors to describe human-environment interactions. They span a range of disciplines such as anthropology, economics, human geography, political science, and sociology.

Improvements in education and public awareness are widely regarded as the solution to environmental problems. Courses are now being offered at graduate level to prepare and train students to be passionate thinkers who solve some of the world's most complex environmental issues. The course also prepares students for careers in government, environmental law, environmental management (including coastal and marine management), environmental health sciences (including nutrition, agriculture, and horticulture), humanities, research and teaching, psychology, and the nonprofit sector. Others are environmental architecture and environmental design, environmental engineering, environmental science and sustainability, marine sciences, and renewable energies. Everyone who desires to save the world should take these courses.

Careers in environmental social sciences are growing. This is due in large part to increased public awareness about the seriousness environmental issues. Businesses, governments, and educational institutions are all beginning to understand that they must act to solve these problems [17]. More information on environmental social sciences is available in the books in [6, 18-33] and the following journals devoted to it:

- *Journal of Environmental and Social Sciences*
- *Asia Proceedings of Social Sciences*
- *Journal of Political Ecology*

- *Journal of Environmental Economics and Management*
- *Journal of Environmental Psychology*
- *Journal of Environmental Assessment Policy and Management*
- *Journal of Occupational and Environmental Medicine*
- *International Journal of Environmental Research and Public Health*
- *American Journal of Sociology*
- *Journal of Industrial Ecology*
- *International Journal of Occupational and Environmental Health*
- *Journal of World-Systems Research*
- *Energy & Environment*
- *Energy Research & Social Science*
- *Social Science Quarterly*
- *World Development*
- *Environmental Politics*
- *The Social Science Journal*
- *International Sociology*

REFERENCES

[1] M. N. O. Sadiku, Y. P. Akhare, A. Ajayi-Majebi, and S. M. Musa, "Environmental social sciences," *International Journal of Trend in Research and Development,* vol. 7, no. 6, Nov.-Dec. 2020, pp. 52-54.

[2] D. J. Penn, "The evolutionary roots of our environmental problems: Toward a Darwinian ecology," *The Quarterly Review of Biology,* vol. 78, no. 3, September 2003, pp. 275-301.

[3] "Integrating the social sciences with the environmental and earth sciences," December 2016,

http://monkeysuncle.stanford.edu/?p=1537

[4] "Environmental social science," *Wikipedia,* the free encyclopedia

https://en.wikipedia.org/wiki/Environmental_social_science

[5] H. Folmer and O. Johansson-Stenman, "Does environmental economics produce aeroplanes without engines? On the need for an environmental social science," *Environmental Resource Economics*, vol. 48, 2011, pp. 337–361.

[6] J. Hannigan, *Environmental Sociology.* New York: y Routledge, 2nd edition, 2006

[7] T. Forsyth, "Environmental social movements in Thailand: How important is class?" *Asian Journal of Social Sciences,* vol. 29, no. 1, 2001, pp. 35-51.

[8] S. Parry and J. Murphy, "Towards a framework for analysing interactions between social science and environmental policy," *Evidence & Policy,* vol. 9, no. 4, pp. 531-546.

[9] "Natural resources and environmental management,"

https://cms.ctahr.hawaii.edu/nrem/NremNews-Details/join-nrem

[10] Y. Sharma, "Social sciences must respond to environmental change,"

https://www.scidev.net/global/environment/news/social-sciences-must-respond-to-environmental-change.html#:~:text=Social%20sciences%20should%20more%20effectively%20research%20%E2%80%9Chuman%20causes%2C%20vulnerabilities%20and,jointly%20by%20UNESCO%2C%20the%20Organisation

[11] G. Palsson et al., "Reconceptualizing the 'anthropos' in the anthropocene: Integrating the social sciences and humanities in global environmental change research," *Environmental Science & Policy,* vol. 28, April 2013, pp. 3-13.

[12] . G. Martinez, "Environmental Research = Social Sciences + Humanities + Environmental Science," September 2015,

https://www.ecologic.eu/12476

[13] H. Walach, H. J. Mutter, and R. Deth, "Inorganic mercury and Alzheimer's disease—Results of a review and a molecular mechanism," *Diet and Nutrition in Dementia and Cognitive Decline*, 2015, pp. 593-601.

[14] S. Yearley, "Environmental social science and the distinction between resource use and industrial pollution: Reflections on an international comparative study," *Ambiente & Sociedade,* vol. 8, no.1, January/June, 2005.

https://doi.org/10.1590/S1414-753X2005000100002

[15] V. Y. Martin, "Four common problems in environmental social research undertaken by natural scientists," *BioScience,* vol. 70, no.1, January 2020, pp. 13–16.

[16] M. Stern, (2018). "Social science theory for environmental sustainability: A practical guide," 2018,

https://my.usgs.gov/hd/publications/social-science-theory-environmental-sustainability-practical-guide

[17] "Environmental studies vs. Environmental science: What is the difference?"

August 2018,

https://unity.edu/uncategorized/environmental-studies-vs-environmental-science/

[18] E. F. Moran, *Environmental Social Science: Human–Environment Interactions and Sustainability.* John Wiley & Sons, 2010.

[19] R. A. Hinde, *Individuals, Relationships and Culture: Links Between Ethology and the Social Sciences.* Cambridge, UK: Cambridge University Press, 1987.

[20] T. O'Riordan (ed.), *Environmental Science for Environmental Management.* Routledge, 2014.

[21] Ö. Bodin and C. Prell, *Social Networks and Natural Resource Management: Uncovering the Social Fabric of Environmental Governance.* Cambridge, UK: Cambridge University Press, 2011.

[22] I. Vaccaro, E. A. Smith, and S. Aswani (eds.), *Environmental Social Sciences: Methods and Research Design.* Cambridge University Press 2010.

[23] *Decision Making for the Environment: Social and Behavioral Science Research Priorities.* National Academies Press, 2005.

[24] J. Y. Zhang and M. Barr, *Green Politics in China: Environmental Governance and State-society Relations.* Pluto Press, 2013.

[25] F Berkhout, M. Leach, and I. Scoones, *Negotiating Environmental Change: New Perspectives from Social Science.* Edward Eglar, 2003.

[26] M. Gray, J. Coates, and T. Hetherington (eds.), *Environmental Social Work* Routledge, 2013.

[27] G. Spaargaren, A. P. J. Mol, and F. H. Buttel (eds.), *Governing Environmental Flows: Global Challenges to Social Theory.* Cambridge, MA: The MIT Press, 2006.

[28] F. Berkhout, *Negotiating Environmental Change: New Perspectives from Social Science.* Edward Elgar, 2003.

[29] D. M. McAllister, *Evaluation in Environmental Planning: Assessing Environmental, Social, Economic, and Political Trade-Offs.* Cambridge, MA: The MIT Press, 1983.

[30] A. V. Kneese and B. T. Bower, *Environmental Quality Analysis: Theory & Method In The Social Sciences.* The Johns Hopkins University Press, 2013.

[31] J. Y. Zhang and M. Barr, *Green Politics in China: Environmental Governance and State-society Relations.* Pluto Press, 2013.

[32] J. Lockyer and J. R. Veteto, *Environmental Anthropology Engaging Ecotopia: Bioregionalism, Permaculture, and Ecovillages.* Berghahn Books, 2013.

[33] R. Haining, *Spatial Data Analysis in the Social and Environmental Sciences.* Cambridge, UK: Cambridge University Press, 1993.

CHAPTER 4

ENVIRONMENTAL EDUCATION

"*The world has achieved brilliance without wisdom, power without conscience. Ours is a world of nuclear giants and ethical infants.*" – Anonymous

4.1 INTRODUCTION

Environment is everything around us. All life is interconnected by the environment, which is the place where organisms interact with each other. Everything that surrounds us and on which our life depends is our environment. Our room, our home, our office, our village or city, our family and friends, the land, air, drink, sunshine, and rain are all part of our environment. A person's environment constitutes the events and culture that the person lived in. Negative balance between nature and human will have a negative impact on the environmental carrying capacity.

Since industrial revolution, people largely destroyed the environment for economic and industrial development. Environmental conditions around the world has been deteriorating at an alarming rate. Human activities have brought problems on the environment such as overpopulation, pollution, burning fossil fuels for energy, global warming, habitat conservation, endangered species, water scarcity, food scarcity, malnutrition, and loss of biodiversity.

The health of the environment is connected with humans' well-being and economic prosperity. The environment sustains all life on earth and our economy thrives on a healthy environment. Over recent years, rapid industrialization, urbanization, and exploitation of resources have seriously affected the environment. The environment adversely degrades due to environmental mismanagement and there are imbalance and disharmony. Oppressed communities are in general disproportionately affected by environmental degradation, which in turn causes poverty and violence [1].

Education cannot remain insulated from contemporary social and environmental issues.

The ultimate goal of education at any level is shaping human behavior. It is the main factor in preventing and resolving environmental problems. The idea of environmental education (EE) was born out of the realization that solution to complex local and global environmental problems cannot be accomplished by politicians and experts alone, but requires the collective participation of an informed public [2].

This chapter provides an introduction on environmental education. It begins by discussing what environmental education is all about. It provides the goals and objectives of environmental education. It presents the principles of environmental education. It covers the strategies for achieving environmental education. It mentions the roles and responsibilities of environmental educators. It considers how environmental education is practiced in different nations. It highlights the benefits and challenges of environmental education. The last section concludes with some comments.

4.2 CONCEPT OF ENVIRONMENTAL EDUCATION

Environmental education refers to a process that allows individuals to explore environmental issues, engage in problem solving, and take action to improve the environment. It may be regarded as efforts in teaching environmental issues and how individuals and businesses can change their behavior in order to achieve sustainable existence. It is provided to help people understand the world around them and know how to take care of it properly so that the world can be a better place [3]. It provides important opportunities for students to become engaged in important global environmental and related issues. It aims for a democratic society in which environmentally literate citizens participate with creativity and responsibility. Its objectives include awareness, knowledge, attitude, skills, and participation. It teaches individuals how to weigh various sides of an environmental issue through critical thinking.

Environmental education continues to grow as an active area of inquiry within the field of education. To be successful, environmental education must be carried on at all academic levels by all types of educational institutions. Environmental education can be considered in the past, the present, and the future. Environmental education emerged in the 1970s in response to the growing awareness and global concern about human damage to the environment. The fields of outdoor education and conservation education set the stage from which EE emerged.

Environmental education entered the public discourse in the late 1960s. The National Environmental Education Act of 1990 was a federal mandate to encourage states in the US to develop EE plans. In 1980, the International Environmental Education Program of UNESCO published Strategies for Developing an Environmental Education Curriculum. The goal of environmental education is to create an environmentally literate society.

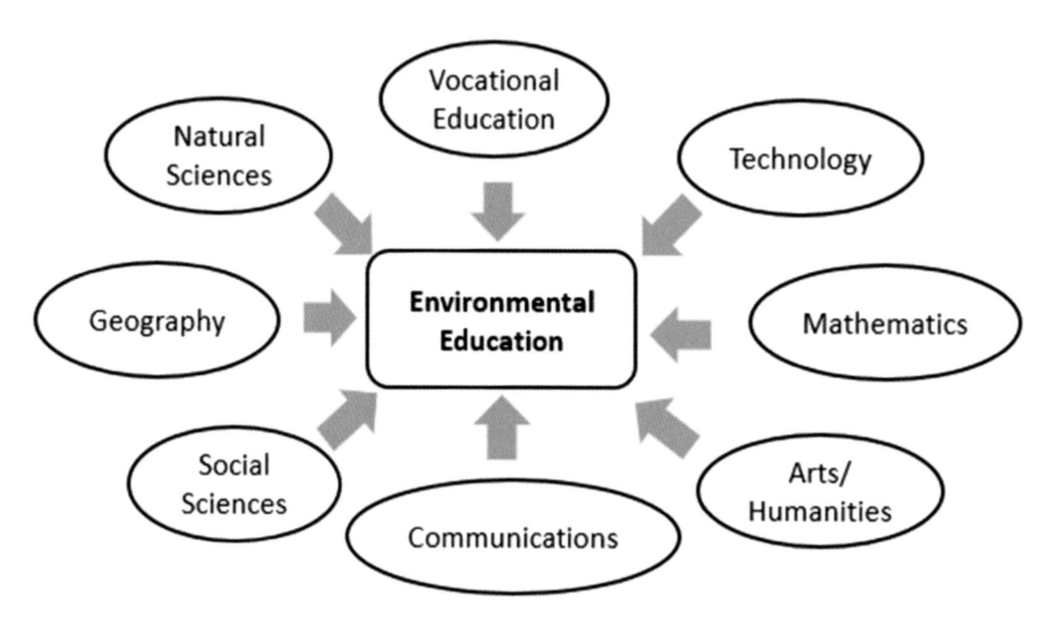

Figure 4.1 The multidisciplinary nature of environmental education [4].

Environmental education is a learning process that increases people's awareness and knowledge about the environment and the necessary skills and expertise to address the challenges. As shown in Figure 4.1, EE is a multidisciplinary field embracing disciplines such as science, social sciences, technology, communication, mathematics, and geography [4]. The primary goal of environmental education is to change environmental behavior through increasing environmental knowledge. It aims at developing a world population that is aware and concerned about the current environmental problems and aims at the prevention of new ones. Consciously or unconsciously, culture invades our teaching practice in environmental education.

Environmental education focuses on [5]:

1. Engaging with citizens of all demographics.
2. Thinking critically, ethically, and creatively when evaluating environmental issues.
3. Making educated judgments about those environmental issues.
4. Developing skills and a commitment to act independently and collectively to sustain and enhance the environment.
5. Enhancing their appreciation of the environment, resulting in positive environmental behavioral change.

4.3 GOALS AND OBJECTIVES OF ENVIRONMENTAL EDUCATION

Environmental education seeks to educate the public about environmental issues. It is usually focused on elementary and high school students, and with the members of the public. The First Intergovernmental Conference on EE was held in Tbilisi, Georgia, in October 1977. The conference developed the Tbilisi goals and objectives for EE. The goals of environmental education are [6]:

- To foster clear awareness of, and concern about, economic, social, political, and ecological interdependence in urban and rural areas.
- To provide every person with opportunities to acquire the knowledge, values, attitudes, commitment, and skills needed to protect and improve the environment.
- To create new patterns of behavior of individuals, groups, and society as a whole towards the environment.

The categories of environmental education objectives are:

- *Awareness*—to help social groups and individuals acquire an awareness and sensitivity to the total environment and its allied problems.
- *Knowledge*—to help social groups and individuals gain a variety of experience in, and acquire a basic understanding of, the environment and its associated problems.
- *Attitudes*—to help social groups and individuals acquire a set of values and feelings of concern for the environment and the motivation for actively participating in environmental improvement and protection.
- *Skills*—to help social groups and individuals acquire the skills for identifying and solving environmental problems.
- *Participation*—to provide social groups and individuals with an opportunity to be actively involved at all levels in working toward resolution of environmental problems.

4.4 PRINCIPLES OF ENVIRONMENTAL EDUCATION

Environmental education strongly believes that humans can live compatibly with nature. To achieve the objectives stated above, a set of guiding principles for environmental educators have been developed. These are [7, 8]:

1. Education should emphasize our interdependence with other peoples, other species, and the planet as a whole.
2. Education should help students move from awareness to knowledge to action.
3. Teachers, students, and schools in the world's richer countries should reduce their consumption of the world's resources.

4. Students must have opportunities to develop a personal connection with nature.
5. Education should be future-oriented.
6. We must relearn "old wisdoms" from native peoples to re-connect to the planet.
7. Teachers should incorporate media literacy into every school subject.
8. Teachers should be facilitators.
9. Teachers should be good role models for their students and "walk their talk."
10. Environment should be considered in its totality (natural, artificial, technological, ecological, moral, aesthetic).
11. Must focus on current, potential environmental situations.
12. Active participation in prevention and control of pollution.
13. Examine root cause of environmental degradation.
14. Provide an opportunity for making decisions and accepting their consequences.

Best environmental education practices are derived from these EE principles. The principles have been researched, critiqued, revisited, and expanded as EE has evolved.

4.5 ACHIEVING ENVIRONMENTAL EDUCATION

Environmental education includes both formal and informal education and training that prepares the individual to become able to balance between his vital needs and the natural environment. Figure 4.2 shows different components of environmental education [9], which includes environmental awareness, environmental challenges, environmental knowledge, environmental literacy, environmental science, understanding, attitudes, action, ecology, and motivation. In order to achieve EE goals, the following strategies are necessary [10, 11].

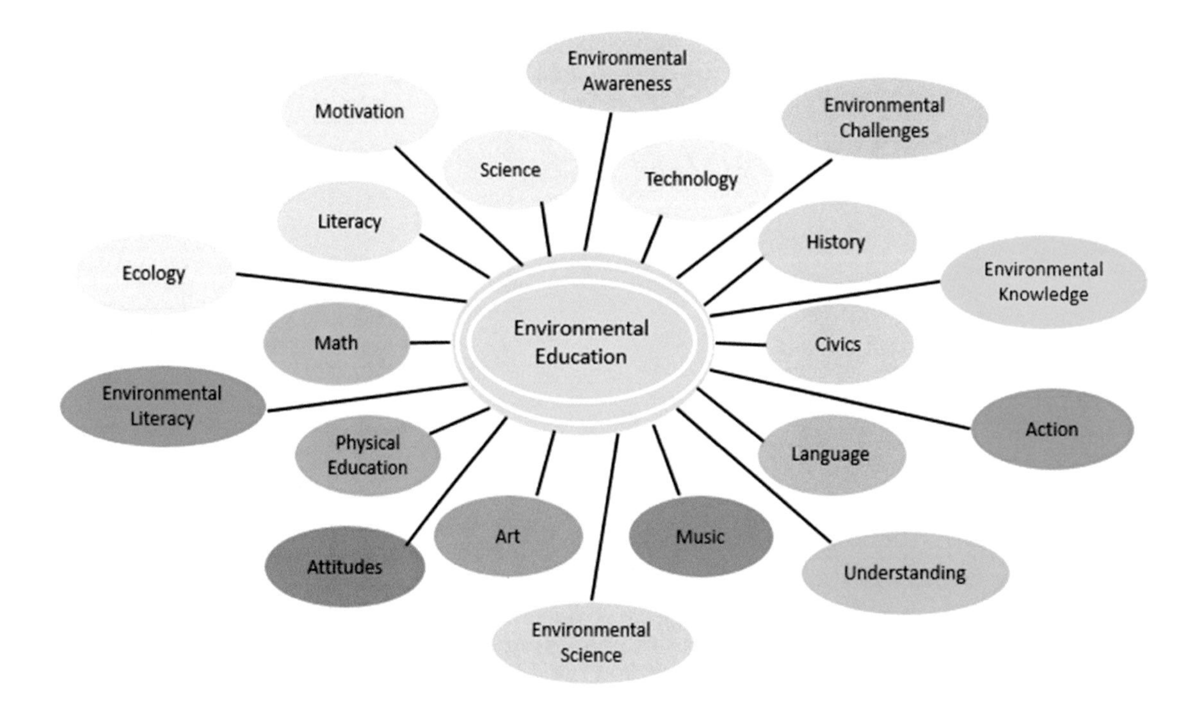

Figure 4.2 Components of environmental education [9].

1. *Environmental Awareness:* The primary aim of environmental awareness is for people to become environmentally aware and be responsible in managing and protecting the environment. Promoting public environmental awareness is crucial goal in contemporary education. Environmental education awareness affects one's character, changes attitudes, and shapes how one feels and believes about the environment. Promoting environmental awareness helps us to become better environmental stewards and participate in creating a brighter future for our children. It empowers individuals, groups, institutions, and organizations to properly explore environmental issues and make plans that are environmentally friendly, savvy, and beneficial. Environmental education teaches people how to think, not what to think. It encourages individuals to take an active role in environmental problem-solving. It prepares the future generation for a green society. Figure 4.3 shows the importance of access to environmental education [12]. Means of raising public awareness include media (newspapers, radio, TV), parks and gardens, seminars, workshops, conferences, awareness materials (brochures, posters, videos, etc.), exhibitions, public awareness events, festivals, websites, and other Internet-based tools. Socio-scientific and science issues can also help raise environmental awareness. The deficiency of environmental awareness jeopardizes the long-term health and security of animals, plants and humans. Environmental awareness can be promoted at schools and colleges by introducing the 3 R's: reduce waste, reuse resources, and recycle materials. Schools can teach students why trees are important to the environment and then organize tree planting days at school. Schools can also encourage children to switch off all appliances and lights when not in use in order to conserve energy.

Figure 4.3 The importance of access to environmental education [12].

2. *Environmental Literacy:* This is an important social goal. Environmental literacy depends on a personal motivation to help ensure environmental quality as well as quality of life. Environmental educators have a unique privilege to produce an environmentally literate citizens, who are empowered with the knowledge, skills, and motivation to tackle the environmental challenges that we face today and in the future and are poised to take action on pressing environmental issues. A well-informed public will formulate opinions and make choices that will improve the quality of the environment and reduce anthropogenic stresses on ecosystems. Environmental

literacy could be systematically achieved by introducing environmental issues and information into the educational curriculum at all levels of the educational system. Environmental Literacy is the desired outcome of environmental education which strives to provide learners with: sound scientific information, skills for critical thinking, creative and strategic problem solving, and decision-making. Creating environmental awareness and promoting education for environmental literacy are the means to ensure that humans do not degrade environment but conserve it for the future.

3. *Environmental Legislation:* Environmental legislation is a collection of many laws and regulations aimed at protecting the environment from harmful actions (for instance the National Environmental Policy Act (NEPA). Political action is a critical part of environmental education. Local implementation of EE could be considered. Government regulation makes certain behavior mandatory or prohibited. Government can have companies pay lower prices for environmentally more benign goods. Economic incentives could be in forms of subsidies, which stimulate desired behavior, and taxes, which discourage undesired behavior.
4. *Environmental Policy:* Environmental policy is aimed at balancing environmental protection and the conservation of natural resources with other policy goals, such as affordable energy and economic growth. Environmental policy in the United States involves governmental actions at the federal, state, and local level to protect the environment and conserve natural resources. Environmental policy can include laws and policies addressing water and air pollution, chemical and oil spills, smog, drinking water quality, land conservation and management, and wildlife protection, such as the protection of endangered species.
5. *Environmental Protection:* Environmental protection is the practice of protecting the natural environment by individuals, organizations, and governments. It is important to protect and conserve our environment in order to maintain its overall health. These are the simple things we can do to help protect the earth:
   - Reduce, reuse, and recycle: Cut down on what you throw away. Follow the three "R's" to conserve natural resources and landfill space.
   - Volunteer: Volunteer for cleanups in your community. You can get involved in protecting your watershed, too.
   - Educate: When you further your own education, you can help others understand the importance and value of our natural resources.
   - Conserve water: The less water you use, the less runoff and wastewater that eventually end up in the ocean.
   - Choose sustainable: Learn how to make smart seafood choices at www.fishwatch.gov.
   - Shop wisely: Buy less plastic and bring a reusable shopping bag.
   - Use long-lasting light bulbs: Energy efficient light bulbs reduce greenhouse gas emissions. Also flip the light switch off when you leave the room.
   - Plant a tree: Trees provide food and oxygen. They help save energy, clean the air, and help combat climate change.
   - Do not send chemicals into our waterways. Choose non-toxic chemicals in the home and office.
   - Bike more. Drive less.

4.6 ENVIRONMENTAL EDUCATORS

Since the 1970s, non-governmental organizations that focused on environmental education continued to grow, while the number of teachers implementing environmental education in their classrooms has increased. Careers in EE generally require discovering and planning how to resolve environmental issues and challenges facing us today. Environmental educators seek to educate the public about nature and environmental issues. They work along with businesses, schools, nature reserves, nonprofit organizations, and other groups to raise awareness of environmental issues. They can work as federal government park ranger, outdoor education teacher, conservation educators, environmental scientist, environmental engineer, and teachers in academic institutions. Environmental education professionals may be employed by oil companies, utility companies, and other companies that have a large impact on the environment.

Figure 4.4 illustrates environmental education in curriculum process [13].

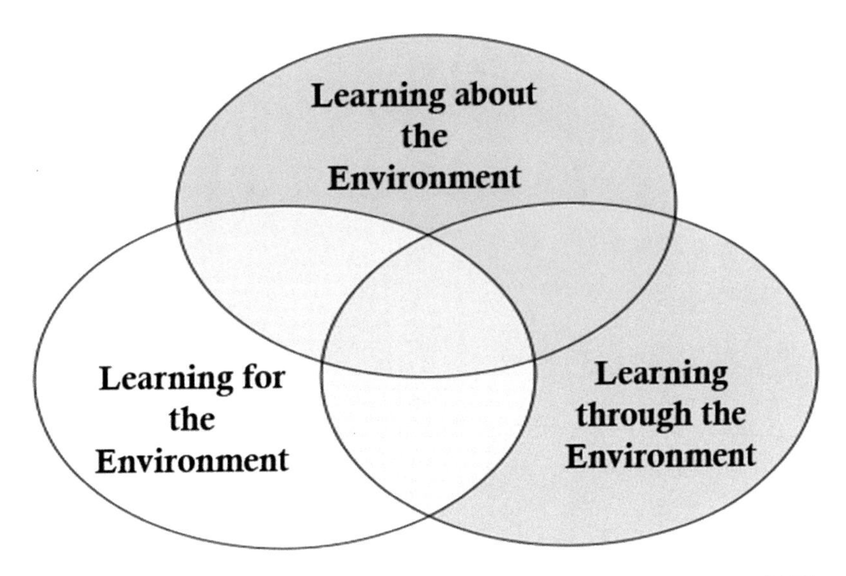

Figure 4.4 Environmental education in curriculum process [13].

Although environmental educators have diverse responsibilities which may vary from job to job, the following list includes typical duties expected of them [14]:

- Analyze and interpret data obtained from literature reviews, research, and sample findings, imagery, and computer model predictive data
- Research, interview, and teach about leading scientists and theories
- Explain and illustrate how the environmental event or trend in question may impact the earth, and human or animal populations
- Communicate lessons to stakeholders on environmental situations or environmental trends according to the scope of the course. Examples may include climate change, watershed protection, or recycling and composting
- Develop environmental awareness: engage members of the public and stakeholders with developed materials for environmental education and training.
- Develop curriculum-based resources and provide support for other educators.
- Write resource material for personal use, other educators, or take-home materials for students and stakeholders, that includes paper, online, oral lecture and multimedia
- Facilitate discussions between differing interests in order to enrich course offerings with cross-disciplinary perspectives and understanding
- Develop feedback plans and programs
- Mentor junior team members and educators

Environmental educators plan events to educate the next generation about environmental issues. Before promoting environmental awareness in your community, you must first educate yourself using environmental news, books, Internet, online courses, articles, videos, and other resources. Familiarize yourself with environmental issues affecting your own community. Start and focus on one environmental issue at a time, perhaps the one that strikes you as the most urgent. Popular environmental problems include ozone depletion, desertification, deforestation, mining, oil drilling, production of plastic goods, loss of biodiversity, etc. Figure 4.5 shows some environmental educators at work [15].

Figure 4.5 Some environmental educators at work [15].

4.7 GLOBAL ENVIRONMENTAL EDUCATION

Environmental education is concerned with the understanding of environments at local, national, and global levels. It is a process that assists individuals, communities, and organizations to learn more about the natural and built environments, and develop skills on how to address global challenges. Foundation for Environmental Education (FEE) is the world's largest environmental education organization, with members in 77 countries [16]. The Global Environmental Education Partnership (GEEP) is a global partnership committed to advancing environmental literacy to create a more just and sustainable future through the power of education.

Today, the concept of the environmental education has been accepted in many countries around the world. We will now consider how environmental education is practiced in some nations.

- *United States:* Environmental education programs for children, college students, and adults continued to develop in the 1990s to the current day. In 1990, the US Congress passed the National Environmental Education Act, which placed the Office of Environmental Education under Environmental Protection Agency (EPA). EPA's definition of EE is as follows: "Environmental education is a process that allows individuals to explore environmental issues, engage in problem solving, and take action to improve the environment. As a result, individuals develop a deeper understanding of environmental issues and have the skills to make informed and responsible decisions." EPA supports environmental education, professional development, youth and educator recognition, and grants. The EPA also has a list of the components of what should be gained from EE [5].

- Awareness and sensitivity to the environment and environmental challenges
- Knowledge and understanding of the environment and environmental challenges
- Attitudes of concern for the environment and motivation to improve or maintain environmental quality
- Skills to identify and help resolve environmental challenges
- Participation in activities that lead to the resolution of environmental challenges.

Environmental education as an educational strategy is prevalent across the US. The North American Association for Environmental Educators develops mechanisms both to strengthen standards for environmental education and to make it achievable.

- *Indonesia:* The natural resources of this nation are being exploited at a rapid rate. The environmental education and awareness among the populace is poor. The shortcomings in environment education in Indonesia include underqualified teachers, the civil service mentality, the still-pervasive chalk-and-talk pedagogy, and the effect of the examination system. The youths are not currently being exposed to effective environmental education. Perhaps the most culturally appropriate way forward in Indonesia is to frame pro-environment behavior and responsibility as a form of citizenship, and specifically that environmental education should be taught as a separate subject [17].
- *United Kingdom:* Environmental education policy in the UK makes scant reference to nature and the issue of our underlying attitude towards it. Such policy is somewhat pre-occupied with the issue of meeting "sustainably" what are taken to be present and future human needs. The current UK National Curriculum policy of delivering environmental education through traditional subjects has some shortcomings. While the attitudes of children towards nature and the environment are generally positive, they can involve a number of limitations, dichotomies, and ambivalences which their education could help them address [18].
- *Nigeria:* Environmental education will enable youths to have a greater voice on environmental issue if effectively implemented in Nigeria. It has been observed that the relationship between the students' knowledge and attitude towards the environment is a negative, little or no relationship. Nigeria as well as other African nations have started to realize the role of environmental education in solving environmental issues. Nigerian government proposed various strategies to curb environmental issues such as loss of biological diversity, threat to food security, destruction due to flooding, soil erosion, desertification, and deforestation [19].
- *Canada:* In 2002, the Canadian government developed a broad vision for environmental learning. The Ministries of Education have introduced a variety of environmental education initiatives into school settings. A noteworthy educational initiative is referred to as Environmental Studies Programs (ESPs). ESP participation greatly impacts secondary school students attitudes toward the environment due to the fact that the secondary school programs are interdisciplinary. Another successful environmental education initiative is integrated ESP, an approach where environmental issues are integrated into an interdisciplinary curriculum at the secondary school level [20]. There are few post-secondary educational programs in Canada in Aboriginal languages.
- *Sweden:* Environmental education in the Swedish schools has been around since the 1960s. The Swedish curriculum states the responsibilities of schools to prepare students with knowledge of the conditions for good environment stewardship. Every individual can understand his or her impact on the environment through education. One is encouraged to become an environmental citizen, which implies that one assumes the challenge as an individual of making changes in one's daily life in favor of a better environment [21].

- *South Africa:* The international organizations such as the United Nations influenced South Africa on concerns about human rights and accountability. The United Nations Commission on Sustainable Development holds the Johannesburg Summit in Johannesburg, South Africa. The summit brought together tens of thousands of participants with ever-increasing demands for food, water, shelter, sanitation, energy, health services, and economic security. The Chair establishes Rhodes as a leading organization for environmental education research and academic training in Africa [22].
- *China:* China's environmental problems continues to attract international concern. Today, the condition of China's water resources is undesirable, the atmospheric environment is being severely polluted, solid waste pollution and noise pollution are increasing, and the ecological environment is degenerating rapidly as a result of reduced biodiversity. Strategic objectives include ensuring safe water supply to the megacities of China by working with farming communities to sustainably irrigate their farms without exhausting. Cultural context is an important aspect to defining project success and gaining a cultural license to operate a project within communities. Environmental education plays an important role in China's sustainable development. Many universities use multimedia learning environments instead of traditional face-to-face teaching method. More and more computers are being used in environmental education in China [23,24].
- *Pakistan:* Aga Khan University Institute for Educational Development (AKU-IED) in Pakistan was established realizing the educational needs of developing countries such as Pakistan. The introduction of environmental education as a means to addressing environmental issues presents a major challenge to the dominant conception. Environmental education is critical for developing educators with context-specific learning experiences in a variety of environments [25].
- *Ecuador:* Environmental education is used as a conservation strategy in the buffer zone of southern Ecuador. By bringing political ecology and engaged pedagogy together, one can provide students with the skills needed to confront social and environmental challenges. In this manner, environmental education could be transformed into a powerful strategy to address underlying power structures and worldviews that encourage environmental destruction and social inequality. Organizations and governmental entities have embraced environmental education as an integral part of their agendas [26].
- *Sudan:* Environmental movement started in Sudan in 1975 with a growing awareness of the environmental problems such desertification, deforestation, pollution, and water contamination. This movement led to the establishment of the leading Institute of Environmental Studies (IES) in the University of Khartoum. The IES is the first of its kind in Africa and Middle East. The objectives of IES include encouraging multidisciplinary research in environmental issues and providing broad consultancy task to councils within the Sudan. Environmental education is already infused in education programs at all education levels [27].
- *Kenya:* This nation has demonstrated a major commitment to educating and training its citizens in the preservation and conservation of resources for present and future generations. Because of this commitment, Kenya is one of the leading countries on the African continent in the implementation of environmental education and training programs at all levels of education. In initiating and developing a program of action for environmental education in Africa, UNEP has outlined the following objectives: (1) To provide a framework for catalyzing, coordinating and organizing environmental education and training activities in Africa at the national, regional and sub-regional levels for all systems of education; (2) To provide an agreed upon framework for the programming of environmental education and training activities in Africa, which countries can use to assess and promote the incorporation of environmental dimensions in the education

and training system; (3) To promote general awareness of the need for environmental education and training in the region, based on information exchange about activities, programs and materials which have been developed; (4) To provide a framework to facilitate compilation and production of materials for developing environmental education and training in the region; (5) To create educational conditions which will make people in the Africa region aware of, concerned about, committed to, and equipped with skills necessary for solving environmental problems and preventing new ones [28].

- *Brazil:* Brazil is a country of great dimensions and diversity. In the last four decades, activists, commentators, and critics have witnessed a consolidation of environmental education in Brazil. This may be attributed to the Brazilian environmental movement that emerged during a period of military dictatorship in the 1970s. Transformative and critical forms were promoted most particularly by social movement activists and environmentalists, whose goals of social transformation saw education as one of the chief tools to achieve this. There have also been politically driven efforts to institutionalize environmental education in Brazil [29].
- *Russia:* Although the achievements of Russia in reducing environmental problems still fall short of many Western nations, progress in environmental education in Russia has already been rapid and reasonably successful. There are five basic levels of environmental education and training for the Russian population.

➢ The Preparatory level includes elements of pre-school environmental education.
➢ The General level is environmental education in primary and secondary schools and in non-specialized colleges, technical and higher schools.
➢ The Specialized level is environmentally oriented training in biological, geographical, agricultural, chemical, and other faculties of universities.
➢ The Special level is the training of master's and postgraduate students in environmental specialties at classical and technical universities.
➢ The fifth level is the retraining of experts in the different branches of the environmental sciences (continuing professional development).

Each level has specific requirements and organizational approach of environmental education [30].

- *Poland*: Environmental education in Poland has become not only a necessity but also a duty as a result of numerous international recommendations. Its influences are directed towards the whole of society as part of formal and non-formal education. Due to the range and mass scale of influence, television takes priority in the transfer of information about the environment. Internet use is expensive in Poland and only a few can use it in the process of self-learning and self-improvement. The Environmental Education Centers are active in the field of environmental education. Non-formal environmental education is also organized in the workplace [31].
- *India:* India is challenged by environmental degradation and economic growth amidst the paradoxical coexistence of poverty and affluence in their multifarious dimensions. The ultimate drivers of environmental degradation in India include population growth, inappropriate technology, poverty, pollution, and unplanned urbanization. To address these issues, environmental education for sustainable development (EESD) has emerged as an important approach to encourage students to conserve and protect the natural environment in their neighborhood. Field based education is one component of active environmental education which engages students in the learning process such as discussions, writing, asking and answering questions, and engaging in their own learning [32].

4.8 BENEFITS

Environmental education is vital to the survival of mankind. It deeply believes that humans can live compatibly with nature and act equitably toward each other. It connects us to the world around us and raises our awareness of issues impacting the natural and built environments. It is built on sustainability and on how people and nature can exist in productive harmony. It enables global thinking, cooperative learning, and problem solving skills. It fosters a sense of connection to the natural world and encourages conservation of irreplaceable natural resources. EE is critical for a sustainable future.

Environmental education benefits students, educators, school systems, businesses, organizations, and the larger world. Environment-based education is often lauded by educators as an ideal way to integrate academic disciplines and promote conservation of the natural environment. Such education teaches students how to think, not what to think. They become the greatest agent of change for the long term protection and stewardship of the environment. They also become better environmental citizens. EE empowers individuals, groups, institutions, and organizations to properly explore environmental issues and make plans that are environmentally friendly, savvy, and beneficial. EE constitutes a comprehensive lifelong education and has many benefits for everyone in the society. It can benefit learners of all ages. It is a concept that transcends the classroom. Today, faith communities, such as churches, synagogues, temples, and mosques, are providing venues for adult environmental education.

Other benefits of EE are [33]:

- Solving various environment challenges
- Playing crucial role in town planning
- Providing knowledge related to environmental issues
- Providing various directives for environment
- Imagination and enthusiasm are heightened
- Learning transcends the classroom
- Critical and creative thinking skills are enhanced
- Tolerance and understanding are supported
- State and national learning standards are met for multiple subjects
- Biophobia and nature deficit disorder decline
- Healthy lifestyles are encouraged
- Communities are strengthened
- Responsible action is taken to better the environment
- Students and teachers are empowered
- Enhancing the thought of peoples' experiences
- Enlightening people to understand their environment
- Providing room for better living
- Serving as an engine room in creating awareness
- Providing aesthetic beauty

4.9 CHALLENGES

Environmental education is sometimes regarded as an emotionally charged and debatable issue at the crossroads of science, technology, and society. It has been observed that there are disparities between the theoretical environmental education portrayed in academic literature and the environmental education taught in schools. Although some teachers have positive attitudes toward environmental education, most lacked the commitment to actually teach it. Environmentalism is often seen as simply an encroachment on the free market. The environmental challenge is permanent; it will never go away.

As the field of environmental education evolves, its principles have been researched, critiqued, revisited, and expanded. Critics argue that environmental education has failed because it is not keeping pace with environmental degradation. Teachers are underpaid, undersupported, and overwhelmed by trying to prepare students for standardized tests and teaching them essential skills. Where will they get the time to fit environmental education into an already crowded activity? [34]. Policies and practices of environmental education have somewhat overlooked women through gender blindness.

4.10 CONCLUSION

Environmental education is a process that allows individuals to explore environmental issues, engage in problem solving, and take action to improve the environment. It teaches individuals how to weigh various sides of an environmental issue through critical thinking. It is essentially a comprehensive lifelong education for everyone. It is rooted on ecological thinking and respect for all life on the earth. It is critical for a sustainable future. At the grass root level, EE aims to make individuals and human communities understand the complex nature of the natural and the built environments. The field is rapidly expanding in size and increasingly diverse in nature. EE will make this planet a better place to live in for the present and future generations.

The modern environmental education movement have realized that to be successful, greening initiatives require both grassroots support. Hundreds of academic institutions worldwide have organized efforts to reduce energy use, water consumption, and material flows. They are encouraging students to take an active role in environmental education and stewardship. Environmental education has been considered an additional subject in the elementary school level. Figure 4.6 depicts five ways of teaching children about the environment [35]

Figure 4.6 Five ways of teaching children about the environment [35].

The North American Association for Environmental Education (NAAEE) is a professional association for environmental educators. The inclusion of environmental education in the "Every Student Succeeds Act" (replacing No Child Left Behind) signifies an important step forward for American educators. More information on environmental education is available in the books in [4,17,36-50] and the following journals exclusively devoted to it:

- *Environmental Education*
- *Environmental Education Research*
- *The Journal of Environmental Education*
- *Annual Review of Environmental Education*
- *Applied Environmental Education and Communication*
- *Environment and Behavior*
- *Australian Journal of Environmental Education*
- *Brazilian Journal of Environmental Education Research*
- *Canadian Journal of Environmental Education*
- *South African Journal of Environmental Education*
- *International Research in Geographical & Environmental Education*
- *International Electronic Journal of Environmental Education*
- *International Journal of Environmental & Science Education*
- *International Journal of Early Childhood Environmental Education*
- *One Earth*

REFERENCES

[1] "Environmental education: Aim, principles, and concept,"

https://www.ilearnlot.com/environmental-education-aim-principles-and-concept/55271/

[2] P. Hart and K. Nolan, "A critical analysis of research in environmental education, "*Studies in Science Education,* vol. 34, no.1, 1999, pp. 1-69.

[3] M. N. O. Sadiku, Y. P. Akhare, A. Ajayi-Majebi, and S. M. Musa, "Environmental education: A primer," *International Journal of Trend in Research and Development*, vol. 7, no. 6, Nov.- Dec. 2020, pp. 46-50.

[4] S. Lahiri (ed.), *Environmental Education.* New Delhi, India, Studera Press, p. 398.

[5] "Environmental education," *Wikipedia,* the free encyclopedia,

https://en.wikipedia.org/wiki/Environmental_education

[6] "Tbilisi goals and objectives,"

https://sites.duke.edu/eelandscape/round-i/tbilisi-goals-and-objectives/

[7] T. Grant, "Nine principles for environmental education," Green Teacher Magazine,

http://gilesig.org/29Nine.htm

[8] "Environmental education (EE): Objectives, principles and programmes,"

https://www.biologydiscussion.com/environment/environmental-education-ee-objectives-principles-and-programmes/16779

[9] "Environmental education for New Mexico," https://eeanm.org/

[10] "Environmental education," Unknown Source.

[11] "Protecting our planet starts with you,"

https://oceanservice.noaa.gov/ocean/earthday.html

[12] "The importance of access to environmental education," January 2019,

https://givingcompass.org/article/the-importance-of-access-to-environmental-education/

[13] B. C. Shukla, "Environmental education in school curriculum an overall perspective," October 2016,

http://www.pioneershiksha.com/news/3202-environmental-education-in-school-curriculum-an-overall-perspective.html

[14] "What is an environmental educator?"

https://www.environmentalscience.org/career/environmental-educator

[15] "Environmental-awareness,"

https://www.pachamama.org/environmental-awareness

[16] "Positive change on a global scale,"

https://www.fee.global/

[17] L. Parker and K. Prabawa-Sear, *Environmental Education in Indonesia: Creating Responsible Citizens in the Global South?* London, UK: Routledge, 2019.

[18] M. Bonnett and J. Williams, "Environmental education and primary children's attitudes towards nature and the environment," *Cambridge Journal of Education,* vol. 28, no.1, 1998.

[19] N. I. Erhabor and J. U. Don, "Impact of environmental education on the knowledge and attitude of students towards the environment," *International Journal of Environmental & Science Education*, vol. 11, no. 12, 2016, pp.5367-5375.

[20] M. Breunig, "Food for thought: An analysis of pro-environmental behaviors and food choices in Ontario environmental studies programs," *Canadian Journal of Environmental Education*, vol. 18, 2013, pp. 155-172.

[21] D. Thor and P. Karlsudd, "Teaching and fostering an active environmental awareness design, validation and planning for action-oriented environmental education," *Sustainability,* vol. 12, 2020.

[22] A. Gough and N. Gough, "Environmental education research in Southern Africa: Dilemmas of interpretation,"*Environmental Education Research,* vol. 10, no. 3, 2004, pp. 409-424.

[23] "The importance of environmental education,"

https://www.ohiooceanfoundation.org/blog-ohiooceanfoundation/the-importance-of-

[24] W. Xia et al., "Application of computers in environmental education in China," *Proceedings of 2nd International Conference on Education Technology and Computer,* 2010.

[25] A. Shah and S. Jehangir, "Teaching for quality education in environmental education: Challenges and possibilities," *Quality in Education: Teaching and leadership in Challenging Times,* vol. 2, 2006, pp. 565-579.

[26] K. A. Lynch, "Environmental education and conservation in southern Ecuador:

Constructing an engaged political ecology approach," *Doctoral Dissertation,* University of Florida, 2001.

[27] E. H. A. Mohammed, M. Kidundo, and M. Tagelseed. "Environmental Education and public Awareness," *Workshop on Post Conflict National Plan for Environmental Management*, Khartoum, Sudan July, 2006.

[28] M. Korir-Koech, "Environmental education in Kenya beyond the year 2000,"

Journal of Eastern African Research & Development, vol. 21, 1991, pp. 40-52.

[29] F. T. Thiemann, L. M. de Carvalho, and H. T. de Oliveira, "Environmental education research in Brazil," *Environmental Education Research*, vol. 24, no. 10, 2018, pp. 1441-1446.

[30] N. S. Kasimov, S. M. Malkhazova, and E. P. Romanova, "Environmental

Education for Sustainable Development in Russia," *Journal of Geography in Higher Education*, vol. 29, no.1, 2005, pp. 49-59.

[31] A. M. Wójcik, "Informal environmental education in Poland," *International Research in Geographical & Environmental Education*, vol. 13, no.3, 2004, pp. 291-298.

[32] R. Alexandar and G. Poyyamoli, "The effectiveness of environmental education for sustainable development based on active teaching and learning at high school level-a case study from Puducherry and Cuddalore regions, India," *The Journal of Sustainability Education,* December 2014.

[33] "Top 10 benefits of environmental education,"

https://www.plt.org/educator-tips/top-ten-benefits-environmental-education/

[34] **"**Green failure: What's wrong with environmental education?"

https://www.google.com/search?q=Green+Failure%3A+What%E2%80%99s+Wrong+With+Environmental

+Education%3F&rlz=1C1CHBF_enUS910US910&oq=Green+Failure%3A+What%E2%80%99s+Wrong+With

+Environmental+Education%3F&aqs=chrome..69i57j69i60.2022j0j7&sourceid=chrome&ie=UTF-8

[35] "Benefits of environmental education in kids,"

https://www.iberdrola.com/environment/enviromental-education-for-kids

[36] J. Palmer, *Environmental Education in The 21st Century: Theory, Practice, Progress and Promise.* Routledge, 2002.

[37] J. Palmer and P. Neal, *The Handbook of Environmental Education.* Routledge, 2003.

[38] A. Bodzin, B. S. Klein, and S. Weaver (eds.), *The Inclusion of Environmental Education in Science Teacher Education.* Springer, 2010.

[39] G. Reis and J. Scott (eds.), *International Perspectives on the Theory and Practice of Environmental Education: A Reader*. Springer, 2018.

[40] D. E. Pinn (ed.), *Environmental Education: Perspectives, Challenges and Opportunities.* Nova, 2017.

[41] C. Saylan and D. Blumstein, *The Failure of Environmental Education (And How We Can Fix It).* University of California Press, 2011.

[42] S. Lahiri, *Environmental Education.* Studera Press, 2019.

[43] A. Kaur, *Environment Education.* Twenty First Century, 2017.

[44] A. Russ and M. E. Krasny, *Urban Environmental Education Review.* Comstock Publishing Associates, 2017.

[45]J. E. Otiende, W. P. Ezaza, and R. Boisvert (eds.), *An Introduction to Environmental Education.* University of Nairobi Press, 1991.

[46] A. Stewart, *Developing Place-responsive Pedagogy in Outdoor Environmental Education*. Springer, 2020.

[47] R. B. Stevenson et al. (eds.), *International Handbook of Research on Environmental Education.* Routledge, 2012.

[48] G. A. McBeath et al., *Environmental Education in China.* Edward Elgar Publishing, 2015.

[49] T. Lloro-Bidart and V. Banschbach (eds.), *Animals in Environmental Education: Interdisciplinary Approaches to Curriculum and Pedagogy*. Palgrave Macmillan, 2019.

[50] J. L. Woodhouse and C Knapp (eds.), *Place-Based Curriculum and Instruction: Outdoor and Environmental Education Approaches.* Psychology Press, 2000.

CHAPTER 5

ENVIRONMENTAL ENGINEERING

"*We abuse land because we regard it as a commodity belonging to us. When we see land as a community to which we belong, we may begin to use it with love and respect.*" - Aldo Leopold

5.1 INTRODUCTION

Manufacturing and industry run our society and we cannot do without them. But they create a lot of environmental problems such as pollution and waste. The progress of modern civilization has taken place at the expense of the environment in which we all live. Mother Nature has her intrinsic ways of treating waste and pollution. The only problem is that it will take her a long time to ensure that the pollution levels are acceptable. Consequently, engineers have intervened and found ways in treating the generated wastes [1].

In some countries, environmental degradation is more serious than war or terrorism. Technology is both the source of environmental problems and potential solutions. Environmental engineering is the branch of engineering that focuses on protecting the environment by minimizing and managing waste and pollution [2].

Environmental engineering applies science and engineering principles to develop ways to protect human health and minimize the adverse effects of human activities on the environment. The field emerged in response to widespread public concern about environmental degradation such water and air pollution. It is important for developing ways to protect the environment and efficiently manage natural resources [3]. Figure 5.1 illustrates downtown Los Angeles depicting environmental engineering [4].

Figure 5.1 Downtown Los Angeles depicting environmental engineering [4].

This chapter introduces the reader to environmental engineering. It begins by explaining the concept of environmental engineering. It discusses what environmental engineers do. It provides some applications of environmental engineering. It highlights some benefits and challenges of environmental engineering. The last section concludes with comments.

5.2 CONCEPT OF ENVIRONMENTAL ENGINEERING

Environmental engineering is an engineering process that considers the environment in as many aspects as are thought to be relevant. It strives to use environmental understanding and technology to serve mankind by decreasing production of environmental hazards and the effects of those hazards already present in the soil, water, and air. It creates solutions that will protect human health and improve the quality of the environment. It studies the effect of technological advances on the environment and addresses local/global environmental issues such as global warming, ozone depletion, water pollution, air pollution, climate environment, and electromagnetic environment, environmental law, and waste management [5]. Some of these components are illustrated in Figure 5.2 [6].

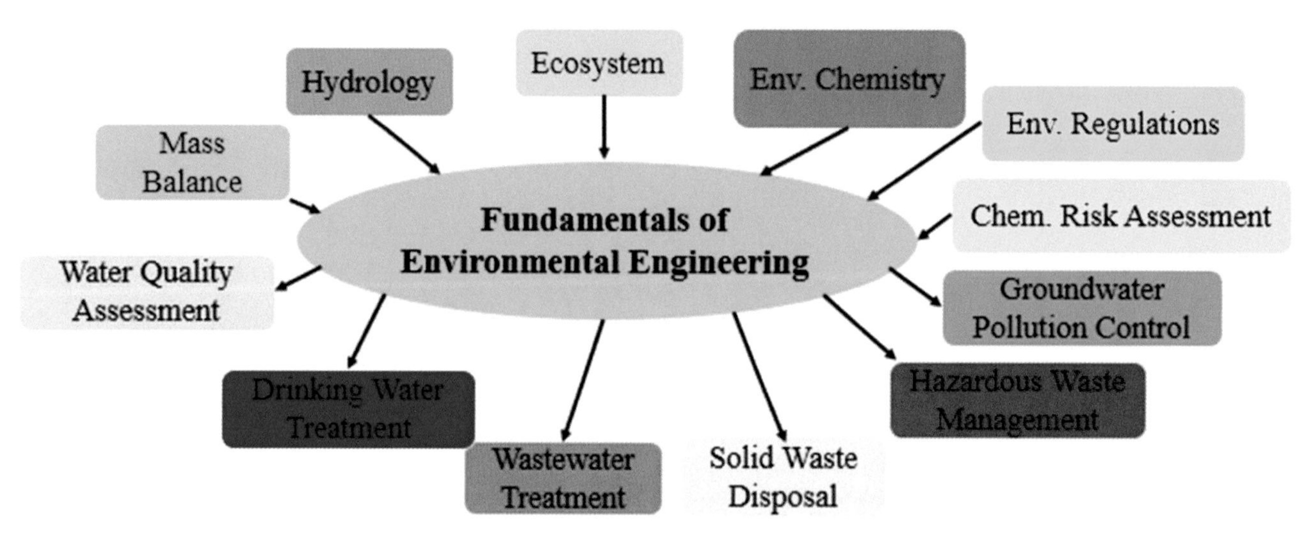

Figure 5.2 The components of environmental engineering [6].

Traditionally, environmental engineering started as part of civil engineering. Most civil engineering programs were renamed "Civil & Environmental Engineering" in many universities across the nation. Today, environmental engineering is a synthesis of various disciplines including civil engineering, chemical engineering, mechanical engineering, chemistry, biology, mathematics, geology, ecology, hydrology, agronomy, medicine, microbiology, economics, architecture, etc. Environmental engineering has broadened to include air pollution, waste management, acoustic and electromagnetic wave contamination, and pollutants migration dynamics. Figure 5.3 compares environmental engineering with civil engineering [7].

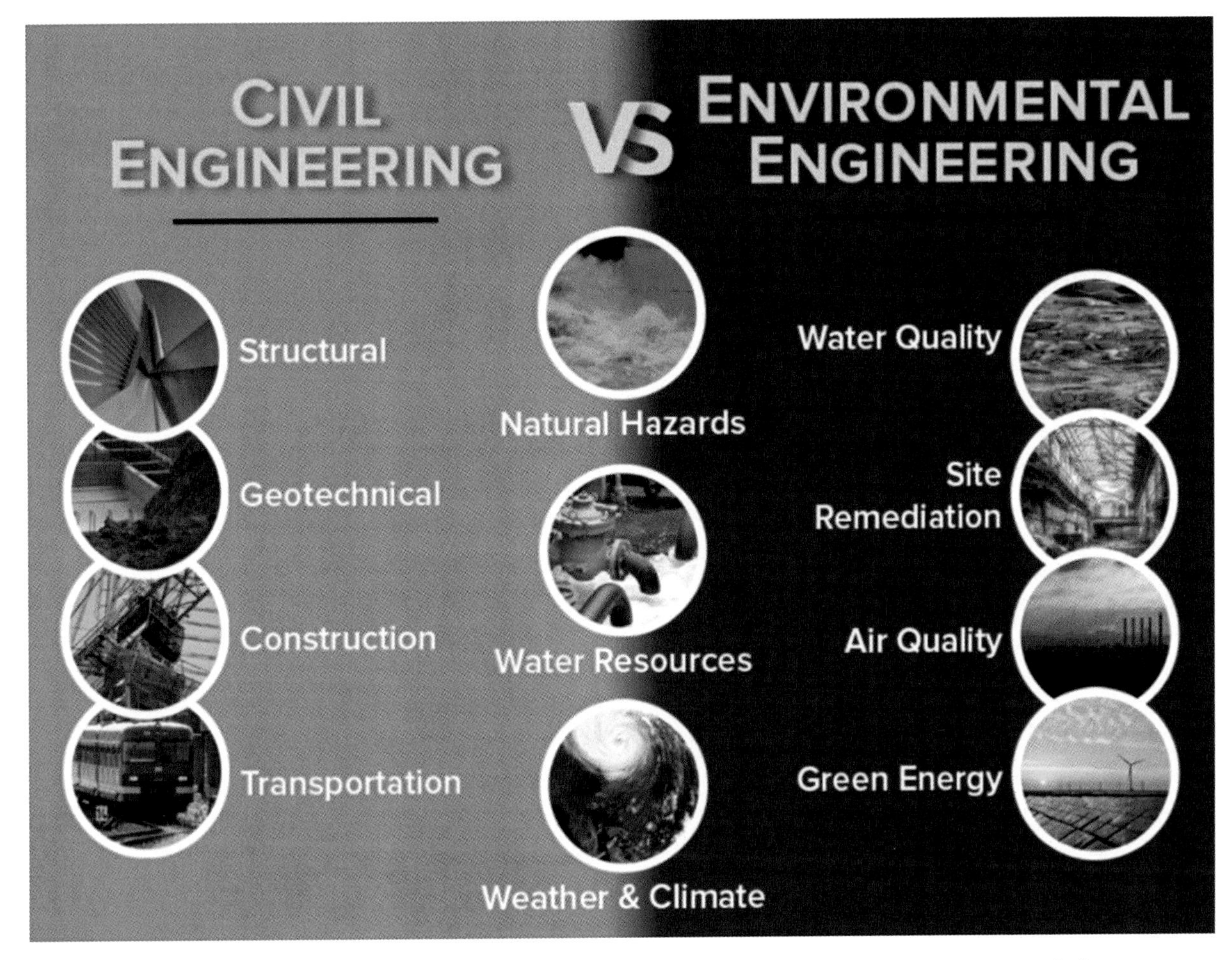

Figure 5.3 Comparing environmental engineering with civil engineering [7].

Environmental engineering jobs are stimulating, since you get to work on a wide range of projects and on multidisciplinary teams consisting of structural, mechanical, electrical, and geotechnical engineers and biologists, chemists, planners, economists, lawyers, and politicians. Projects in environmental engineering involve the treatment and distribution of drinking water, control of air pollution, waste management, marketing environmental-control equipment, and environmental assessments. Mathematical modeling is widely used to evaluate and design systems [8].

5.3 ENVIRONMENTAL ENGINEER

Environmental engineers provide leadership on environmental issues to the community where they live and work. Like other engineers, environmental engineers are problem solvers, innovators, entrepreneurs, and global leaders. They use environmental science and engineering principles to develop solutions to environmental problems. They design, plan, and implement measures to prevent, control, or remediate environmental hazards. They design and build infrastructure for the supply of water, the disposal of waste, and the control of pollution. They design systems such as water distribution systems, waste management systems, and pollution control technology. The duties of an environmental engineer involve water supply, waste water and solid management, air and noise pollution control, environmental sustainability, environmental impact assessment, climate changes, etc. He also works to improve recycling, waste disposal, public health, and water and air pollution control. His major goal is to prevent or reduce undesirable impacts of human activities on the environment. Since he is dealing with public issues, he needs training on public health, public policy, and public education. Due to litigation in environmental issues, environmental engineers must be familiar with applicable laws. He needs skills of integration, communication, and conceptualization in addition to a traditional engineering background.

While jobs do vary from place to place, an environmental engineer is typically responsible for the following [9]:

- Assess industrial, commercial and residential sites for their environmental impact
- Calibrate equipment used for air, water, or soil sampling
- Design systems for waste management, reclamation, transfer and disposal on land, sea, and air
- Advocate best remediative procedures for site clean-up and contamination
- Advise policymakers and companies on relevant issues
- Evaluate the current system performance and incorporate innovations or develop new technologies to enhance environmental protection
- Collect field samples and observations for data and observations
- Investigate environmentally related complaints, recording data and compiling a report based on these
- Ensure that stakeholders are in regulatory compliance for waste management and disposal
- Collect, construct and evaluate environmental impact statements
- Must have good communication skills as a team player
- Communicate with a variety of technical and non-technical stakeholders
- Liaise with interdisciplinary teams for a holistic solution to environmental engineering problems
- Prepare, review, and update environmental investigation reports
- Design projects leading to environmental protection, such as water reclamation facilities, air pollution control systems, and operations that convert waste to energy
- Obtain, update, and maintain plans, permits, and standard operating procedures
- Provide technical support for environmental remediation projects and legal actions
- Analyze scientific data and do quality-control checks
- Monitor progress of environmental improvement programs
- Inspect industrial and municipal facilities and programs to ensure compliance with environmental regulations
- Advise corporations and government agencies about procedures for cleaning up contaminated sites

Environmental engineers are required to have a bachelor's degree in environmental engineering or a related engineering field, such as civil engineering, chemical engineering, or general engineering. It is helpful to gain practical experience since some employers place high value on it. Some employers also require certification as a professional engineer (PE). Ongoing training and education are continuously expected to keep up with advances in technology and current government regulations. A master's degree in environmental engineering or MBA is often required for promotion to management level. It can open opportunities for teaching at higher education institutions or conducting research.

Environmental engineers can work in industrial or in outdoor settings. Industries such as manufacturing, textile processing, petroleum and petrochemical, pharmaceutical, and meat processing employ environmental engineers to ensure compliance with environmental regulations. Outside industry, environmental engineers are employed by national and regional environmental agencies, local health departments, and public works departments. Environmental engineers in government jobs develop regulations to prevent mishaps. They work with lawyers and business owners to address environmental problems and sustainability. Figure 5.4 show what an environmental engineer typically does [10].

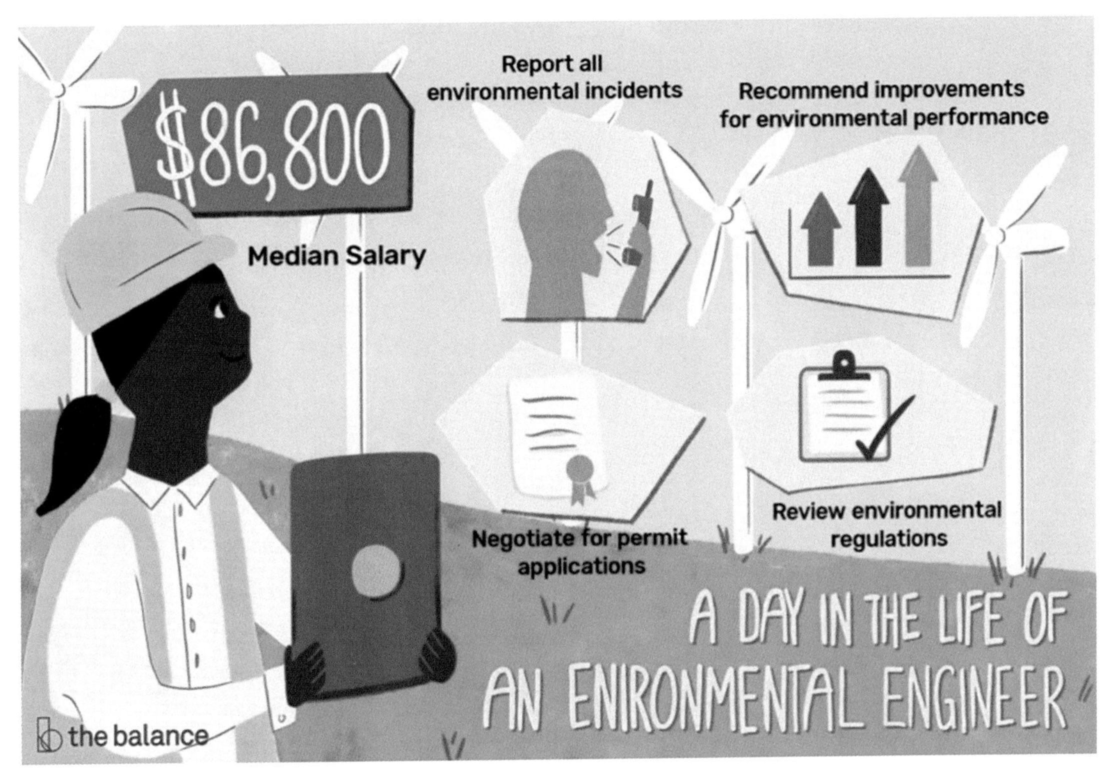

Figure 5.4 What an environmental engineer typically does [10].

5.4 APPLICATIONS

Some subdivisions of environmental engineering include natural resources engineering, ecological engineering, environmental health engineering, and environmental engineering education.

- *Ecological engineering*: Ecological engineering may be defined as the design of sustainable ecosystems that integrate human society with its natural environment for the benefit of both. Its goals include the restoration of ecosystems that have been substantially disturbed by human activities and the development of new sustainable ecosystems that have both human and ecological values. This must focus on creating things reinforced by their changing contexts. Ecological engineering must be flexible as to take on a looping character that updates the system to meet changing requirements. Rather than fighting against a context change, ecologist work with it, admit ignorance, knowing that every realization itself changes the context from its original state to a certain degree [11].
- *Environmental Management System* (EMS): An EMS is a set of processes and practices that enable an organization to reduce its negative environmental impacts while increasing its operating efficiency. EMS is a structured framework under which an entity can manage environmental impacts. This is a useful tool for organizations that intend to integrate environment management in their corporate policy. It is a tool to measure environmental performance to validate continuous improvement. EMSs such as IS014001 and EMAS may be major growth areas for the accredited certification. The voluntary adoption of EMS by companies has become a vital supplement to mandatory environmental policies.
- *Product Environmental Engineering:* This plays an important role in improving the quality and reliability of products. It includes environmental condition determination, prediction of environmental worthiness, environmental testing and evaluation technologies and environmental

engineering management techniques. The systemic application of product environmental engineering in every phase of materiel development could save money [12]. Reliability is closely related to environmental worthiness. All of them are the properties of quality. Environmental test and reliability test are carried out and coordinated in the processes of product development.

- *Environmental Health Engineering* (EHE): Environmental health engineering deals with solving environmental problems that have measurably adverse public health impacts. Their work includes improving the quality of recycling, waste disposal, public health, water and air pollution control, and the detection of pollutants and tracking them back to their source. EHE applies principles of engineering, soil science, biology and chemistry in order to create solutions to the management of many environmental problems facing mankind. Another key function of environmental engineers is the They can work with companies to find ways to reduce pollutants.
- *Environmental Engineering Education* (EEE): Environmental awareness is growing along with the mounting global environmental problems. Efforts are being made to re-orient environmental engineering education to promote the concept of sustainability as the primary goal of environmental management. Environmental education covers both general (life, earth, and economic) and technological education areas that are related to the environment. The goal of environmental education is professional training of the future environmental engineers [13]. The general public also needs environmental awareness. If every citizen becomes exposed to the subject of the "environment," their ignorance toward the state of the environment will slowly disappear.
- *Green design:* Engineering design requires the use and generation of different materials. The concept of green design is design for the environment. It refers to practices that are intended to yield products with little or no environmental impact. For example, design engineers apply the concept whenever designing a plant to minimize pollution, to save money, and to reduce wastes. It stresses on recycling and reusing materials. This new idea is catching on like wildfire as industries, academia, and government are implementing it as much as possible [1].

More applications of environmental engineering include contaminated soil remediation, risk assessment, minerals processing, environmental regulation development, environmental value engineering, waste management, environmental health and safety, natural resource management, noise pollution, and air pollution.

5.5 BENEFITS AND CHALLENGES

Environmental engineering blends the best aspects of art and design with the sciences of physics, geology, ecology, and chemistry. Environmental engineers apply the principles of engineering and natural sciences to develop solutions to environmental problems. Environmental engineers work to improve recycling, waste disposal, public health, and water and air pollution control. They work to prevent, control or remediate any hazards to the environment using their engineering expertise. Their work may deal with issues such as waste disposal, erosion, and water and air pollution. They provide safe water supplies for societal and environmental needs, reduce emissions of pollutants, manage and remediate sites, and protect public health. They often collaborate with business people, government officials, scientists, and legal professionals toward a common goal such as determining ways to eliminate or reduce pollutants [14].

The environmental problems faced today has no boundaries. Some analysts argue that the damage caused to the global environment by each incremental emission of CO is very small and insignificant. It is challenging for environmental engineers to detect the presence of air pollutants and track them back to their source. Ocean pollution can present even greater challenges in identifying the source.

5.6 CONCLUSION

Environmental engineering is a relatively new discipline, which combines science and engineering principles to improve the quality of the environment. An increasingly health-conscious society is eager to find environmental engineers who can prevent problems rather than simply control those that already exist. This requires that environmental engineers must be proactive in developing alternative solutions. Environmental problems such as water and air pollution, waste generation, and excessive land use are increasingly threatening life-support systems. This requires engineering to allow economic development without environmental mismanagement.

The demand for environmental engineers continues to increase. The future of environmental engineering seems boundless because it is a discipline that will be required for almost all future technologies in energy conservation. More information on environmental engineering can be found in numerous books in [15-55]. One may also consult the following related journals:

- *Journal of Environmental Engineering*
- *Journal of Civil, Construction and Environmental Engineering*
- *Journal of Environmental Health Science and Engineering*
- *Environmental Engineering and Management Journal*
- *American Journal of Environmental Science and Engineering*
- *Iranian Journal of Environmental Health Science & Engineering.*
- *Case Studies in Chemical and Environmental Engineering*
- *International Journal of Environmental Engineering*

REFERENCE

[1] T. Baccay, "Environmental engineering across the engineering curriculum; Classroom case studies and database application, Part I," *Masters Thesis,* Albert Nerken School of Engineering Environmental, October 15, 1996.

[2] "Environmental engineering," *Wikipedia*, the free encyclopedia

https://en.wikipedia.org/wiki/Environmental_engineering

[3] M. N. O. Sadiku, O. D. Olaleye, and S. M. Musa, "Environmental engineering: A primer," *International Journal of Trend in Research and Development,* vol. 6, no. 3, May- Jun. 2019, pp. 102-104.

[4] B. Barragan, "Downtown LA added 7,551 apartments in the last six years" June 2017,

https://la.curbed.com/2017/6/6/15750352/downtown-la-new-apartment-construction

[5] M. M. Mahamud-López and J. M. Menéndez-Aguado, "Environmental engineering in mining engineering education," *European Journal of Engineering Education,* vol. 30, no. 3, 2005, pp. 329-339.

[6] A. D. Kney, "CE 321 Introduction to Environmental Engineering and Science," https://sites.lafayette.edu/kneya/courses/courses-taught/ce-321/

[7] "What is civil and environmental engineering?"

https://cee.engr.uconn.edu/about-us/what-is-cee

[8] J. A. Nathanson, "Environmental engineering,"

https://www.britannica.com/technology/environmental-engineering

[9] "What is an environmental engineer?"

https://www.environmentalscience.org/career/environmental-engineer

[10] D. R. Mckay, "What does an environmental engineer do?" June 2019,

https://www.thebalancecareers.com/environmental-engineer-526013

[11] T.F.H. Allena, M. Giampietrob, and A.M. Little, "Distinguishing ecological engineering from environmental engineering," *Ecological Engineering*, vol. 20, 2003, pp. 389–407.

[12] X. G. Li and S. P. Zhang, "The systemic application of product environmental engineering for materiel development," *Proceedings of the 8th International Conference on Reliability, Maintainability and Safety*, 2009, pp. 374 – 377.

[13] N. E. A. Basri et al., "Introduction to environmental engineering: A problem-based learning approach to enhance environmental awareness among civil engineering students," *Procedia - Social and Behavioral Sciences,* vol. 60, 2012, pp. 36 – 41.

[14] J. Lucas, "What is environmental engineering?" October 2014,

https://www.livescience.com/48390-environmental-engineering.html

[15] P. O. Fanger, *Thermal Comfort: Analysis and Applications in Environmental Engineering.* Copenhagen, Denmark: Danish Technical Press, 1970.

[16] M. L. Davis and D. A. Cornwell, *Introduction to Environmental Engineering.*

McGraw-Hill, 5th ed., 2013.

[17] R. Weiner and R. Matthews (eds.), *Environmental Engineering*. Butterworth-Heinemann, 4th Edition, 2003.

[18] J. A. Salvato, N. L. Nemerow, and F. J. Agardy, *Environmental Engineering*. Hoboken, NJ: John Wiley & Sons, 5th ed., 2003.

[19] S. Franzie, B. Markert, and S. Wunschmann, *Introduction to Environmental Engineering.* Weinhei, Germany: Wiley-VCH Verlag, 2012.

[20] M. Davis and S. Masten, *Principles of Environmental Engineering and Science.* McGraw-Hill, 3rd ed, 2013.

[21] J. A. Nathanson, *Basic Environmental Technology*. Prentice Hall, 2000.

[22] R. O. Mines, *Environmental Engineering: Principles and Practice.* Wiley-Blackwell, 2014.

[23] N. Gaurina-Medjimurec (ed.), *Handbook of Research on Advancements in Environmental Engineering.* Engineering Science Reference, 2015.

[24] J. C. Small, *Geomechanics in Soil, Rocks, and Environmental Engineering.* Boca Raton, FL: CRC Press, 2016.

[25] F. R. Spellman, *Handbook of Environmental Engineering.* Boca Raton, FL: CRC Press, 2016.

[26] R. E. Weiner and R. A. Matthews, *Environmental Engineering.* 2003, Burlington, M: Elsevier Science, 4th edition, 2003.

[27] M. H. Rahman and A. Al-Muyeed, *Water and Environmental Engineering.* Dhaka, Bangladesh: ITN-BUET, 2012.

[28] C. A. Peters, *Environmental Engineering Science*. Mary Ann Liebert, Inc., Publishers, 2019.

[29] W. Z. Tang and M. Sillanpää, *Sustainable Environmental Engineering.* John Wiley & Sons, 2018.

[30] J. Kuo (ed.), *Air Pollution Control Engineering for Environmental Engineers*. Boca Raton, FL: CRC Press, 2016.

[31] D. Reible, *Fundamentals of Environmental Engineering.* Boca Raton, FL: CRC Press, 2019.

[32] M. Rathinasamy et al. (eds.), *Water Resources and Environmental Engineering II: Climate and Environment.* Springer, 2019.

[33] S. A. Abbasi and T. Abbasi, *Current Concerns in Environmental Engineering.* Nova Science Publishers, 2018.

[34] N. L. Nemerow, F. J. Agardy, and J. A. Salvato (eds.), *Environmental Engineering* (3 Volume Set). John Wiley & Sons; 6th edition, 2009.

[35] J. Zhao et al. (eds.), *Rock Mechanics in Civil and Environmental Engineering.* Boca Raton, FL: CRC Press, 2010.

[36] C. C. Lee, *Environmental Engineering Dictionary.* Anham, 4th edition, 2005

[37] G. M. Masters and W. P. Ela, *Introduction to Environmental Engineering and Science.* Pearson, 3rd edition, 2007.

[38] M. L. Davis and D. A. Cornwell, *Introduction to Environmental Engineering.*

McGraw-Hill Higher Education, 1999.

[39] G. M. Masters and W. P. Ela, *Introduction to Environmental Engineering and Science* Pearson Education Limited,2013.

[40] M. Davis and S. Masten, *Principles of Environmental Engineering and Science.*

McGraw-Hill, Third Edition, 2013.

[41] J. R. Mihelcic and J. B. Zimmerman ·(eds.), *Environmental Engineering: Fundamentals, Sustainability, Design.* John Wiley & Sons, 2014.

[42] A. Chadwick, J. Morfett, and M. Borthwick, *Hydraulics in Civil and Environmental Engineering.* Boca Raton, FL: CRC Press, 5th edition, 2010.

[43] M. J. McPherson, *Subsurface Ventilation and Environmental Engineering.* Springer, 2012.

[44] C. S. Revelle, E. Whitlatch, and J. Wright, *Civil and Environmental Systems Engineering.* Pearson Education Limited, 2013.

[45] C. C Lee, *Handbook of Environmental Engineering Calculations.* McGraw-Hill, 2000

[46] J. A. Salvato, N. L. Nemerow, and F. J. Agardy, *Environmental Engineering.* Wiley, 2003.

[47] R. Weiner, R. Matthews, and P. A. Vesilind, *Environmental Engineering.* Elsevier Science, 2003.

[48] P. A. Vesilind, J. J. Peirce, and R. F. Weiner, *Environmental Engineering.* Elsevier Science, 3rd edition, 2013.

[49] M. J. McPherson, *Subsurface Ventilation and Environmental Engineering.* Chapman & Hall, 1993.

[50] A. Chadwick et al., *Hydraulics in Civil and Environmental Engineering.* London, CRC Press, 5th ed., 2013.

[51] W. R. Niessen, *Combustion and Incineration Processes: Applications in Environmental Engineering.* Boca Raton, FL: CRC Press, 3rd ed., 2002.

[52] R. C. Gaur, *Basic Environmental Engineering.* New Delhi, India: New Age International Publishers, 2008.

[53] M. Kutz (ed.), *Handbook of Environmental Engineering.* John Wiley & Sons, 2018.

[54] S. Franzle, B. Markert, and S. Wunschmann, *Introduction to Environmental Engineering.* Wiley, 2012.

[55] R. A. Corbitt, *Standard Handbook of Environmental Engineering.* McGraw-Hill, 2nd ed., 2004.

CHAPTER 6

ENVIRONMENTAL POLLUTION

"Progress is impossible without change, and those who cannot change their minds cannot change anything." - George Bernard Shaw

6.1 INTRODUCTION

Due to advances in technology and the rapid pace of industrialization, there has been an exponential increase in the number of industries around the world. Although these industries provide employment and meet the needs of our modern society, they have caused complex and serious problems to the environment. Their reckless consumption of natural resources has been mainly responsible for aggravating the problem of pollution, also known as environmental pollution.

Pollution is all around us. It is one of the most serious challenges we face worldwide. It is due to unsustainable anthropogenic activities, resulting in substantial public health problems. ("Anthropogenic" refers to environmental change caused or influenced by people, either directly or indirectly). It is causing an unimaginable and irreplaceable damages to the environment. It is often due to the addition of toxic heavy metals in the air, water, and land that reduce the ability of the contaminated environment to support life. Exposures to environmental pollution remain a major source of health risk throughout the world. The risks are higher in developing countries due to poverty and lack of investment in modern technology [1].

Figure 6.1 Examples of environmental pollution [3].

Although the industrial revolution was a great success in terms of technology and the provision of multiple services, it also introduced the production of massive quantities of pollutants emitted into the air. Today, pollution is occurring on an unprecedented scale around the world. Without a doubt, there is a critical concentration of pollution that an ecosystem can tolerate without being destroyed [2]. Figure 6.1 shows some examples of environmental pollution [3].

This chapter provides an introduction to environmental pollution. It begins by explaining what environmental pollution is all about. It discusses different types of pollution. It identifies some causes of environmental pollution and the various effects it has. It addresses various ways environmental pollution can be prevented. It addresses environmental pollution as a global problem. The last section closes with comments.

6.2 CONCEPT OF ENVIRONMENTAL POLLUTION

The relationship between man and his environment is experiencing profound changes due to modern scientific and technological developments. The environmental deterioration is commonly known as pollution. Pollution of land, air, and water has been of major concern for many years. Environmental pollution now constitutes one of the biggest hazards to both human existence and the existence of nature itself. Environmental pollution is the unfavorable alteration of our surroundings [4]. It is the deliberate or inadvertent addition of impurities or pollutants into the natural environment that causes detrimental effects to natural resources and mankind. It occurs because the natural environment does not know how to decompose the unnaturally generated elements. It can be harmful and dangerous even when it is invisible. It is negatively affecting human beings as well as every living organism on the planet like animals, plants, and aquatic organisms. It is a global problem, affecting both developed as well as developing nations. As illustrated in Figure 6.2, the rapid pace of globalization, urbanization, and industrialization has led to some environmental problems [5].

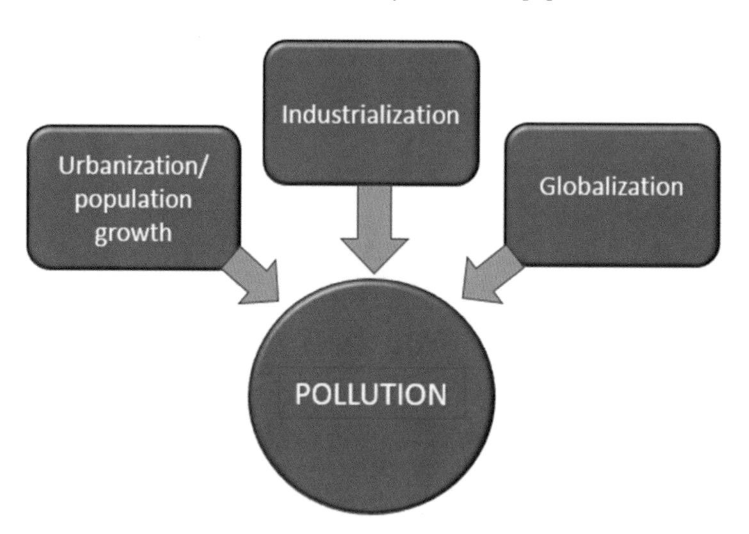

Figure 6.2 The fundamental drivers of pollution [5].

Pollutants are the substances that cause pollution in the environment. They are introduced into the air by several human activities. A pollutant can be any chemical (toxic metal, gases, etc.), biological product, or physical substance (heat, radiation, sound waves, electromagnetic waves, etc.) that is released intentionally or inadvertently by man into the environment. It is hard to attribute a specific health impact to a particular pollutant since populations are exposed to a mix of different pollutants. Environmental pollution generally alters the composition of water, air, and soil of the environment. It is also associated with adverse health effects experienced in the short or long term [6].

As the world's population continues to increase, pollution in all forms becomes an ever-growing challenge. Whether consciously or unconsciously, human activities affect the environment adversely. Environmental pollution may be costly. Therefore, pollution should be dramatically reduced because it is destroying the environment we live in, contaminating the air we breathe, negatively affecting our food and water, and causing diseases in humans. To effectively reduce, minimize, or eliminate pollution requires the joint efforts of both the government and the citizens.

6.3 TYPES OF POLLUTION

Pollution comes from many sources with various consequences. It is essentially the introduction of harmful chemicals into the environment. It can be classified by its physical nature, its source, its recipient, or effects. So there are different types of pollution ranging from physical to economic, political, social, and religious. Here we consider eleven main types of pollution: air, water, land, noise, thermal, radiation, light, urban, plastic, radioactive substances, and oil [4-10].

1. *Air Pollution:* This is the main type of pollution that threatens the environment, humans, plants, animals, and all living organisms. It is capable of causing harm to living organisms. Around the world, nine out of ten people breathe unhealthy air. This makes air pollution the biggest environmental risk for early death, responsible for more than 6 million premature deaths each year. It involves introducing harmful or poisonous substance into the air. It is the contamination of the air by the discharge of detrimental substances. Indoor pollution sources include gases like carbon monoxide, household chemicals, and tobacco smoke. Outdoor air pollutants include benzene, sulfur monoxide, nitrogen dioxide, ozone, and hydrochloric acid from industrial operations. Transportation or vehicle pollution is a major contributor to air pollution, especially in urban areas. Agricultural activities may have a significant impact on air quality.
2. *Water Pollution*: This is contamination of water bodies such as rivers, streams, oceans, seas, lakes, etc. This is an environmental degradation due to introducing pollutants in water bodies. It is caused by a variety of human activities such as domestic, industrial, and agricultural. Water pollution destroys countless species of plants and animals which depend upon water for their survival. It also affects the water quality and makes it unhealthy for use. It can also cause the disruption of food chain and economic downturns. Polluted drinking water causes waterborne diseases such as giardiasis, hookworm, typhoid, liver, and kidney damage. For example, excess of nitrate in drinking water is harmful for infants and human health. The United Nations Educational, Scientific and Cultural Organization (UNESCO) reports on municipal and industrial wastewater contamination to world water supplies in various nations. An example of water pollution is shown in Figure 6.3 [11].

Figure 6.3 A typical example of water pollution [11].

3. *Land/Soil Pollution*: This is land degradation caused by man-made chemicals introduced to the soil. It is due to the addition of unwanted substances to the soil which negatively affect physical, chemical, and biological properties of the soil. It reduces soil productivity and crop yields. The accumulation of solid and liquid waste materials contaminates groundwater and soil. Use of fertilizers, pesticides, deforestation, and pollution due to urbanization causes soil pollution. For example, the use of insecticides and pesticides absorbs the nitrogen compounds from the soil rendering it useless for plants to derive nutrition from.
4. *Noise Pollution:* This involves unwanted sound (or noise) which may be annoying or harmful. Noise is one of the most pervasive pollutant. By definition, noise is unwanted sound or sound without value. On daily basis, we are constantly exposed to heavy noise produced by emergency sirens, loud music, aircrafts, and industrial machinery. For example, you experience sound pollution when you are in a noisy restaurant and cannot hear your friend because everyone else is speaking too loud. Noise can cause irritation, loss of temper, decrease in work efficiency, or loss of hearing. It is, therefore, expedient to control noise. The World Health Organization (WHO) has prescribed optimum noise level as 45 dB by day and 35 dB by night. As a general rule, any sound louder than 80 dB is regarded as hazardous.
5. *Thermal/Heat Pollution:* Thermal pollution is the rise in the temperature of water bodies that is injurious to aquatic life. It is harmful to release heated liquid into a body of water. It is also harmful to release heated air as a waste product of a business into the air. Several industries utilize lot of water for cooling purposes and discharge the used hot water into rivers, streams or oceans. This is thermal pollution. Power plants are usually located on rivers so they can use the water as a coolant. Discharge of hot water from power plants adversely affects feeding in fishes, increases their metabolism, decreases their swimming efficiency, and affects their growth. Thermal pollution reduces biological diversity.
6. *Radiation Pollution:* Radiation is a form of energy travelling through space. This is the increase in the natural background radiation. Radiations can be categorized into two groups, depending on the energy of the radiated particles: the non-ionizing radiations and the ionizing radiations. Non-ionizing radiations are constituted by the electromagnetic waves at the longer wavelength

of the spectrum ranging from near infra-red rays to radio waves. The main effect of non-ionizing radiation on living tissue has only recently been investigated and found to the rise of body tissue temperature. For example, in a microwave oven, the radiation causes water molecules to vibrate faster and thus raise its temperature. Ionizing radiation causes ionization of atoms and molecules of the medium through which it passes. It creates high-speed electrons in a material and breaks chemical bonds. This radiation can be a health hazard. Also, higher levels of ultraviolet radiation can lead to health and environmental effects.

7. *Light Pollution:* This is caused by the obtrusive, misdirected, inappropriate, and excessive use of electric lights to light up the streets. It is the huge amount of light that is produced daily by urban and highly populated areas. While lights help us to see at night, too many lights cause light pollution blocking out the night sky. Light pollution can also be harmful to nocturnal animals. Pollution introduced by light at night is becoming a global problem, especially in urban areas.
8. *Urban Pollution:* The presence of untreated human waste in concentrated areas made the cities the primary sources of pollution. Cities contribute pollution from many sources, including factory smokestacks and wastewater, car exhaust, liquid leaking out of landfills, and sewage treatment plant leakages. Urban pollution has been solved or diminished in the developed nations and it can be solved in the developing nations. Although urban areas are usually more polluted than the rural areas, pollution can spread to remote places where few or no people live.
9. *Plastic Pollution:* This involves the accumulation of plastic products in the environment. The invention of plastic has created one of the most problematic pollution problem ever encountered by mankind and the problem will outlive us for many centuries. For example, the plastic single-use shopping bags have been a major environmental concern. Plastic pollution adversely affects humans and wildlife habitat. Discarding plastic products rapidly fills up landfills and often clog drains. They usually do not biodegrade naturally and add to humanity's mounting litter problem. The plastic creates health problems for the animals since they cannot break down plastic in their digestive system and will usually die from the obstruction. The World Wide Fund for Nature reported some 1.5 million tons of plastic waste from the water bottling industry alone. The overconsumption of resources and the creation of plastics are causing a global crisis of waste disposal. An example of plastic pollution is shown in Figure 6.4 [12].

Figure 6.4 An example of plastic pollution [12].

10. *Radioactive Pollution:* This is due to radioactivity, which is a phenomenon of emission of alpha, beta, and gamma rays due to the disintegration of atomic nuclei of some elements. Human activities can release radiations from activities with radioactive materials (such as mining, handling and processing of radioactive materials, handling and storage of radioactive waste, as well as the use of radioactive reactions to generate energy (nuclear power plants), along with the use of radiation in medicine (e.g. X-rays) and research). The health risk is higher at radiations of higher energy levels, such as gamma-rays. Radioactive Pollution, therefore, may be defined as the increase in the natural radiation levels caused by human activities. Although radioactive pollution is rare, it is highly dangerous, extremely detrimental, and even deadly, when it occurs. It occurs when radioactive substances are released into the environment due to a nuclear explosion, nuclear weapon testing, or disposal of nuclear waste. Improper disposal of radioactive materials can cause dangerous chemicals to be introduced into the environment. Depending on the degree of exposure, radioactive radiation may cause DNA damage to humans.
11. *Oil Pollution:* Oil and gas companies contribute to oil pollution on oceans. Ocean pollution is regarded as a major threat to both marine life and ocean ecosystems. Petroleum products are mined from the earth deep below the ocean surfaces. Oil can end up polluting oceans in many ways. Ships carrying oil have caused devastating oil spills. It is very difficult to clean up mass oil pollution after it has occurred. There is a number of organizations that are dedicated to preventing ocean pollution and cleaning up pollution that has already occurred.

Other types of pollution include personal pollution, nutrient pollution, littering pollution, pharmaceutical pollution, visual pollution, agricultural pollution, transportation pollution, construction and demolition pollution, groundwater pollution, ocean pollution, solid-waste pollution, atmospheric and marine pollution, microbial decaying-process pollution, etc. All types of pollution are interconnected since everything on our planet is interconnected.

6.4 CAUSES OF POLLUTION

Pollution comes from both natural and human-made sources. It is caused by household activities, industrial modernization, factories, agriculture, transport, urban-industrial and technological revolution, rapid exploitation of natural resources, increased rate of exchange of matter and energy, and ever-increasing industrial wastes, urban effluents, and consumer goods. It is related to economic production, modern technology, life-styles, the sizes of human and animal populations, urbanization, industrialization, globalization, etc. Globally human-made pollutants from combustion, construction, mining, agriculture, and warfare are increasingly significant.

- *Humans:* Human activities such as domestic, industrial, and agricultural are mainly responsible for environmental pollution. Humans are primarily responsible for water pollution. The original source of pollution is people or animals already infected with the micro-organisms concerned.
- *Domestic Sources:* These causes of pollution include restrooms, latrines and wastewater from kitchens and bathrooms. Open defecation obviously releases human waste into the environment, which can then be washed into rivers and other surface waters. Solid wastes from households and also from shops, markets, and businesses include food waste, packaging materials and other forms of rubbish.
- *Chemicals:* Several industrial activities release many chemicals into the air every day, thus polluting it. Smog, soot, fumes, smoke from the chimneys, burning of wood, and toxic substances

are the most prevalent air pollutants. Methane, carbon dioxide, sulphur dioxide, and nitrogen dioxide, when released in the air contribute to pollution.

- *Mining:* This is part of the general problem of rapid exploitation of natural resources. Some environmental consequences of mining coal deposits are deforestation, land damage, water pollution, air pollution, noise pollution, ground vibration, and rock dispersal. Such environmental impacts are increasing day by day because the scale of individual mining operations is increasing as mining of lower grade deposits increases. For example, gold mining is just one of world's most pressing global pollution problems. Mercury from gold mining can end up in tuna and poison children. Mining can also contribute to water pollution.
- *Urbanization:* Exodus of people from rural areas to urban centers are responsible for rapid rate of exploitation of natural resources and several types of environment pollution. Urbanization, industrialization, globalization, and rapid economic development have led to increase in energy consumption and waste discharges. Increasing urbanization implies increase in the concentration of human population in limited space which leads to increase in the number of buildings, roads, sewage, and transportation (cars, trucks, buses, motor cycles, etc.), factories, urban wastes, etc. which increase environmental pollution.
- *Climate Change:* Climate change is taking place. It is caused largely by human activities, and poses significant risks for— and in many cases is already affecting a broad range of human and natural systems. Climate change impacts are already manifesting themselves and imposing losses and costs. The risks to public health and the environment from climate change are substantial and far-reaching. Scientists warn that carbon pollution and resulting climate change are expected to lead to more intense hurricanes and storms, heavier and more frequent flooding, increased drought, and more severe wildfires events that can cause deaths, injuries, and billions of dollars of damage to property and the nation's infrastructure.
- *Modern Technology*: The nature of modern technology is closely related to the environmental crisis. Since World War II, productive technologies with intense impacts on environment have displaced less destructive ones. Technological innovation can only be successful if it is able to meet the needs of society. It must enable decision makers to make the best decisions possible. For example, increase in the number of motorized vehicles have caused high levels of air and water pollution.

6.5 EFFECTS OF POLLUTION

Pollution is the introduction of harmful materials into the environment. Pollution of all kinds can have serious negative effects on human health, wildlife, and the environment [13,14].

- *Human Health*: Humans have a profoundly substantial and adverse impact on the environment. Our failure to conduct activities like manufacturing, transportation, large-scale fishing, agriculture, and waste disposal in moderation affects our health as well as the land, air, and water. For example, noise pollution causes hearing loss, high blood pressure, stress, and sleep disturbance. Pollution threatens the continuing survival of human societies. It jeopardizes the supply and flow of clean drinking water. It has an adverse effect on the productivity of both indoor and outdoor workers. Pollution can affect humans by destroying their respiratory, cardiovascular, and neurological systems. It may play a vital role in the increase of allergic diseases in developed world. Figure 6.5 illustrates some health problems related to environmental pollution [15].

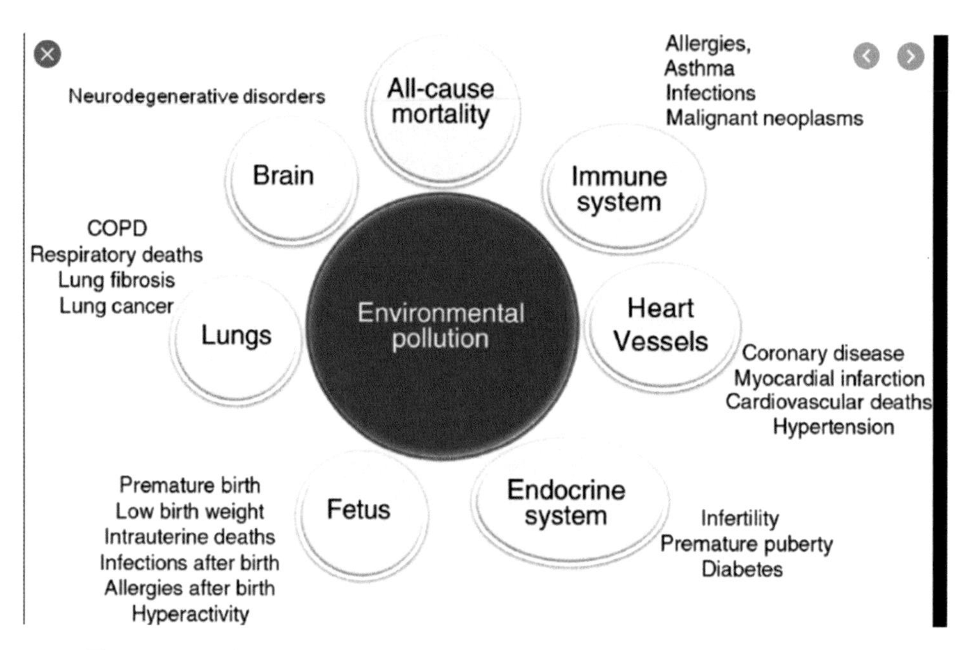

Figure 6.5 Health problems due to environmental pollution [15].

- *Environment:* Environmental management (or pollution control) is necessary to avoid the degradation of the environment due to the accumulation of waste products from overconsumption, heating, agriculture, mining, manufacturing, transportation, and other human activities. Pollution affects the nature, plants, fruits, vegetables, rivers, ponds, forests, animals, etc, upon which they depend on for survival. Pollution of the air, water, and soil is an important threat to human development and also to global sustainable development.
- *Global Warming:* This has become an undisputed fact concerning our current livelihoods. Our planet is warming up and we are definitely responsible for the problem. Pollutants can also have an effect on temperature. The change in temperature can lead to effects such as climate change. High air pollution levels are a major contributor to global warming. The emission of greenhouse gases such CO_2 is leading to global warming. Global warming is causing ice sheets and glaciers to melt. It also contributes to ocean acidification. The risks to public health and the environment from global warming are far-reaching. Climate and weather affect the duration and intensity of disease outbreaks. Mosquito-transmitted parasitic or viral diseases are highly climate-sensitive.
- *Ozone Layer Depletion*: Ozone is poorly soluble and highly reactive gas. The ozone layer is the thin shield high up in the sky (in the stratosphere). It protects life on earth by filtering out harmful ultraviolet radiation (UV) from the sun. It stops ultraviolet rays from reaching the earth. As a result of human activities, chemicals, such as chlorofluorocarbons (CFCs), are released into the atmosphere thereby contributing to the depletion of the ozone layer. Higher levels of ultraviolet radiation reaching earth's surface lead to health and environmental effects such as a greater incidence of skin cancer, cataracts, and impaired immune systems. Nations around the globe are phasing out the production of chemicals that destroy ozone under an international treaty known as the Montreal Protocol. Using a flexible and innovative regulatory approach, the United States already has phased out production of those substances having the greatest potential to deplete the ozone layer under Clean Air provisions enacted to implement the Montreal Protocol.
- *Infertile Land:* The land may become infertile due to continual use of insecticides and pesticides. Various forms of chemicals released into the flowing water affect the quality of the soil and may prevent plants from growing properly.

6.6 PREVENTATION OF POLLUTION

The presence of environmental pollution raises the issue of pollution control or prevention. Great efforts are made to limit the release of harmful substances into the environment. Pollution prevention involves using materials, processes, and practices to minimize or eliminate the creation of pollutants at the source of generation. It include source reduction, increased efficiency, and conservation activities that lead to the reduction in the amount of any pollutant.

Information about environmental pollution can be intimidating and discouraging. The problem appears so huge for an individual to make any difference. As Max Lucado rightly said, "No one can do everything, but everyone can do something." Working together, we can make a difference.

Around the world, government at different levels and individuals should embrace efforts to reduce waste and ensure resource sustainability. They are making efforts to combat pollution through recycling, reusing, waste minimization, mitigating, dilution, and use of technologies. Figure 6.6 shows environmental protection hierarchy [16].

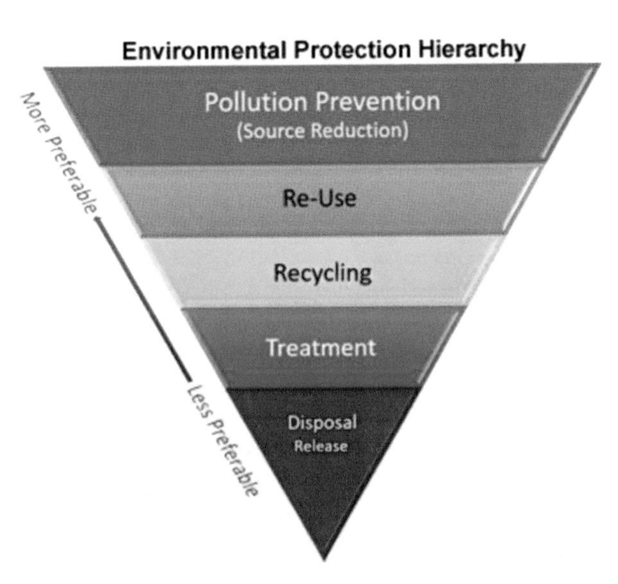

Figure 6.6 Environmental protection hierarchy [16].

- *Waste Disposal*: Waste management is regarded as a service, disposing unwanted materials, typically to landfill. The waste hierarchy minimizes waste by prevention, minimization, reuse, recycling, and recovery before considering landfill. The consumers are expected to dispose product waste and product packaging after the product is consumed. There are three types of waste disposal: residential, commercial, and local government. For example, local government policy typically consists zero waste policy, recycling, and disposal targets [17].
- *Personal Choice*: Plastic shopping bags are used primarily by individuals and the consumption is direct. Since single use bags are a matter of individual choice, people can tackle this problem by simply saying no to plastic bags.
- *Microbial bioremediation:* Many technologies can be used to remove pollutants from soil and water. Microbial bioremediation is a cost-effective and eco-friendly technology that provides sustainable ways to clean up contaminated environments. A wide variety of organisms such as bacteria, fungi, algae, and plants with efficient bioremediating properties have been successfully employed.
- *Environmental Awareness*: Since environmental pollution, directly or indirectly, affects everyone, education and awareness of pollution should be ongoing. Environmental educators and

activists should make effort to increase public awareness of environmental problems. The entire community can be involved through education of the next generation, special television broadcasts, and articles in numerous magazines. This will help to draw the nation's attention to pollution and other threats to the environment. This will also affect public attitudes and behavior toward the environment and encourage individuals to accept some personal responsibility for these problems. Communities across the nation can be urged to observe Earth Day [18]. The protection and improvement of the human environment is a major issue affecting the well-being of people and economic development.

- *Going green:* This can help us come out of the present tough environmental problem. Easy ways to be greener include [19]:

- Implement recycling in the office
- Remember the 3R's: Reduce, Reuse, Recycle
- If recycling already exists, make sure employees are aware
- Educate employees on what can and cannot be recycled (cardboard, plastics, glass fluorescents, and IT equipment
- Purchase recycled paper
- Install water hippos in toilets (a device that sits in the cistern of the toilet and reduces water used with each flush) or use low-flow toilets
- Encourage employees to carpool or bike to work
- Turn off technology and/or appliances when not in use
- Buy LED, CFL (compact fluorescent bulbs) or other long lasting bulbs
- Try to repurpose old technology
- Have heating and air conditioning on when needed, make sure it's not going non-stop
- Go paperless
- Do more things electronically (ex: voting, filing taxes, and various tickets, and meetings)
- Ask boss about possibility of distant working
- Get rid of screen savers and allow products to go into sleep mode
- Order things online or walk / bike, instead of using up fuel
- Try to repair technology instead of replacing it entirely
- Buy a hybrid / electric car
- Use power saving modes for maximum efficiency on all devices

These simple ways can make a significant impact in reducing environmental pollution.

Other ways to reduce pollution include [20]:

1. Environmental planning requires that you evaluate the environmental impact that has to be conducted.
2. Shift to eco-friendly transportation such as electric vehicles and promoting shared mobility (i.e., carpooling, and public transports) could reduce air pollution significantly.
3. Reducing air pollution must involve moving away from fossil fuels, replacing them with sustainable renewables like solar, geothermal, and wind and producing clean energy.
4. Use solar power and wind turbine power, which are both powerful forces against radioactive power and fossil fuel power.
5. Assess the potential environmental impact of an industrial plant by carrying out EIA and warn the industrial plants that do not have EIA.

6. Green building can help solve environmental problems to an extent. The objective of green building is to create environmentally responsible and resource-efficient structures to reduce their carbon footprint.
7. Storage facilities for solid wastes should be built in the city.
8. The wastewater recycling project should be exercised, and a recycling center should be built to reduce water pollution.
9. Environmentally friendly products should be made cheaper to encourage people to use them.
10. Protecting soil, air and water quality should be a fundamental goal of national environmental policy.
11. For prevention of the visual pollution, enough green areas and parks should be built.
12. To prevent noise pollution, the open markets, bazaars, recreation and amusement facilities, schools and parks inside the city should be surrounded by trees and other plants.
13. Need to work on reducing electromagnetic radiation (EMR).
14. Nowadays radiation is a serious issue and the ecological cost of radioactive power plants has become more evident than before.
15. While sanctioning urban land-use, the landscape architects should be engaged by the municipality.
16. The number and quality of the green areas and parks should be increased for fresh and healthy air.
17. The mass media, including print and electronic media and the internet, should be used more intensively to facilitate the transmission of environmental information.
18. Environmental issues should be included in the national education syllables to have well-educated and concerned people about environmental issues.
19. The top-level politicians, executives, administrators and all the other entrepreneurs should also be educated about the environment.
20. Local administrations should be concerned about educating the local people by providing the means (e.g., books, brochures, seminars.) to make them understand the importance of environmental problems.
21. Finally, in order to produce effective solutions for environmental problems, a national approach towards the problems is crucial.

6.7 GLOBAL ENVIRONMENTAL POLLUTION

Environmental pollution is a global problem, affecting both developed as well as developing nations. People and governments around the world are making efforts to combat pollution. In all nations, governments can combat pollution by passing laws that limit the use of chemicals. The rules for protection of the environment are stricter in developed nations than in developing nations. For example, if the environmental standards of developed countries are applied in India, costs will be prohibitive. Thus, the standards selected should be compatible with the country's economic situation. Strict laws should be required in developing nations so that environmental pollution will not result in contamination of the groundwater. Time and scale are the key factors in dealing with pollution anywhere.

The World Health Organization, International Labor Organization, and United Nations Environment Programme provide information on the potential health effects of exposure to environmental pollution.

The Pure Earth is an international non-for-profit organization that is committed to eliminating life-threatening pollution in the developing countries, especially the world's worst polluted places. We now consider how some nations treat environmental pollution.

- *United States:* The awareness of environmental pollution began in the United States when Congress passed the National Environmental Policy Act (NEPA) in the 1970s. Several incidents of pollution helped increase public consciousness. The majority of US citizens are aware of the consequences and dangers of pollution. In the US, to illuminate the streets and parks, about 120 terawatt-hours of energy are consumed. Many nations, including the United States, now severely limit the production of chlorofluorocarbons (CFCs). EPA continues to work with state, local and tribal governments, federal agencies, and stakeholders to reduce air pollution and the damage that it causes. It issues federal emissions standards for common pollutants, thereby helping states to meet the standards.
- *United Kingdom*: Air pollution would continue to be a challenge in England. The 1952 London smog killed between 8,000 and 12,000 people due to immediate effects like bronchitis, pneumonia, and lingering lung damage. Environmental inequity in England varies in relation to: (a) different environmental pollutants, (b) different aspects of socio-economic status, and (c) different geographical scales and contexts. Over the years, London Market insurers have agreed to some 120 environmental pollution settlements with U.S. policyholders. The litigation has not been a happy experience for London insurers. The policyholders themselves have been frustrated in dealing with the London Market [21,22].
- *China:* China has made great economic achievements over the past four decades and has become the second-largest economy in the world. Due to its rapid industrial development and overpopulation, China is one of the Asian countries confronting serious environmental pollution. China is experiencing environmental degradation on a monumental scale and environmental pollution is becoming of major concern in China. The Chinese government has action to address what some have called "one of the greatest environmental threats the earth has ever faced." Processes have been developed to prevent the emission of fluorine. Thermal activation enables leaching of the carbonates at conditions in which the fluorides are unaffected. This allows for the recovery of Ce and the non-fluoride bonded Rare Earth Elements (REE). In view of the abundance of REEs in China, the Chinese REE industry can regard this loss in efficiency an acceptable tradeoff for reducing the environmental impact of the REE extraction processes. It takes just five days for the jet stream to carry heavy air pollution from China to the United States, where it stops clouds from producing rain and snow. Environmental pollution liability insurance was introduced in China in 2006, as part of new approaches for managing environmental risks [23,24].
- *Ethiopia:* Pollution from the industrial sector in this nation has been on the rise, posing a serious problem to the environment. Like industry, agricultural activities are also increasing in Ethiopia, and changing too. Nowadays, agricultural activities in Ethiopia use more pesticides and fertilizers. Fertilizer use in Ethiopia has increased from 140,000 metric tons in the early 1990s to around 650,000 metric tons in 2012. Fertilizer contains phosphate and nitrate and if these reach water bodies, they can cause excessive plant growth [25].
- *India:* About 60 per cent of the air pollution of Delhi (the Indian capital city) is contributed by vehicles. According to the survey report of the National Environmental Research Institute, Nagpur (India) the level of air pollution in Delhi, Calcutta, Bombay, Madras, Ahmedabad, Cochin, Hyderabad, Kanpur, Nagpur, etc. has gone up. Besides industrial wastes from industrial

cities, huge quantity of urban solid wastes also create environmental problems [27]. Pollution is occurring both in urban and rural areas in India due to the fast industrialization, urbanization, and rise in use of motorcycle transportation.

- *Japan:* The fast growth of Japanese economy, due to industrial activities after World War II, has resulted in serious environmental pollution problems. The government often provided assistance to the private sector at the expense of environmental concerns. In 1967, several laws were promulgated by the Japanese government to combat environmental pollution. Despite many pollution control measures taken by administration, Japan faces air, water, and soil contamination problems. The problem of global warming is one to which Japan contributes directly through the production of CO_2 from her homes, cars, and factories. Public acceptance of nuclear energy in Japan has always been a paradoxical and controversial issue [28,29].
- *Nigeria*: Petroleum and petrochemical industries have expanded in Nigeria. As a result, environmental pollution will exacerbate if concerted efforts are not undertaken now to prevent and control oil spills and environmental pollution. Every oil and gas company is required to report spill events in spite of inadequate institutional and legal arrangements for pollution control in the Nigerian oil industry. The working class and the community-based social movements are necessary but not sufficient for the transformation of the Nigerian oil-dependent capitalist economy. The cooperation between the global and local sites of resistance is needed. This will mitigate the excessive accumulation of capital to the detriment of the environment, healthcare, and social condition of the oppressed people of the oil-rich Niger Delta region in Nigeria [30,31].
- *Turkey:* This developing nation has a dynamic economic development and rapid population growth. Western Turkey has cosmopolitan centers of industry, finance, and trade especially in the major cities, whereas the eastern part of the country is relatively underdeveloped. This situation has caused citizens living in the eastern part of country to migrate to the western part of the country, leading to some problems in energy utilization and environmental pollution in the western Turkey. The renewable energy supply in Turkey is dominated by hydropower and biomass, but environmental and scarcity-of-supply concerns have led to a decline in biomass use [26, 32].
- *Canada:* This is no doubt one of the best nations on earth. In spite of this, Canada faces a big issue of pollutants. These pollutants can slowly destroy the high standards which Canada has set consistently over the years. These pollutants would include mercury, nitrogen and sulfur oxides, car emission, and smog. These pollutants remain in Canada in spite of repeated efforts made by the government and other environmental organizations. Canada has a major task in eradicating the pollutants [33]. Canadian pollution and environmental liabilities are increasing risk and exposures for many Canadian businesses. Increasing environmental awareness has led to many recent costly lawsuits for pollution claim, with a devastating cost.

High levels of environmental pollution are reported in Mexico City, Iraq, Rio de Janeiro, Milan, Ankara, Melbourne, Tokyo, and Moscow.

6.8 CONCLUSION

Environmental pollution is an important issue in modern society. It is simply the introduction of contaminants or harmful materials caused by pollutants into the environment. It is a wide-reaching problem that endangers the health and welfare of current and future generations. It is reaching worrying proportions worldwide. In 2015 alone, pollution killed 9 million people worldwide. Advancement in environmental pollution research has provided a wide range of environmental solution techniques [34].

To prepare the future leaders, environmental issues should be included in the education syllables. The mass media, including print and electronic media and the Internet, should be used more intensively to facilitate the transmission of environmental information. Local administrators, politicians, executives, and all the other entrepreneurs should also be educated about the environment. The local people should be made to understand the consequences of environmental pollution.

For more information about environmental pollution, one should consult the books in [35-62] and the following related journals:

- *One Earth*
- *Environmental Pollution,*
- *Science of the Total Environment*
- *Environmental Research*

REFERENCES

[1] M. N. O. Sadiku, P. O. Adebo, A. Ajayi-Majebi, and S. M. Musa, "Environmental pollution: An introduction," *International Journal of Trend in Research and Development*, vol. 7. No. 6, Nov.-Dec. 2020, pp. 155-161.

[2] I. Manisalidis et al., "Environmental and health impacts of air pollution: A review," *Frontiers in Public Health*, vol. 8, no. 14, February 2020.

[3] https://wallsheaven.com/canvas-prints/environmental-pollution-and-green-energy-icon-set-B49816096

[4] K. Kaur,"The effects of environmental pollution on human health," *World Affairs: The Journal of International Issues,* vol. 11, no. 2, Summer 2007, pp. 22-43.

[5] M. Nazeer, U. Tabassum, and S. Alam, "Environmental pollution and sustainable development in developing countries," *The Pakistan Development Review*, vol. 55, no. 4, Winter 2016, pp. 589-604.

[6] V. C. Pandey and V. Singh, "Exploring the potential and opportunities of current tools for removal of hazardous materials from environments," *Phytomanagement of Polluted Sites*, 2019, pp. 501-516.

[7] "Environmental pollution,"

http://nammakpsc.com/wp/wp-content/uploads/2015/08/10.pdf

[8] A.K. Jain, "Environment pollution: Types, causes, effects (PDF download)

November 2019,

https://gradeup.co/environment-pollution-i

[9] "What are the 7 different types of pollution?"

https://examples.yourdictionary.com/what-are-the-7-different-types-of-pollution.html

[10] M. A. Khan and A. M. Ghouri, "Environmental pollution: Its effects on life and its remedies," *International Refereed Research Journal*, vol. 2, no. 2, April 2011, pp. 276-285.

[11] K. Nadeem, "Seven types of pollution,"

https://scientiamag.org/seven-types-of-pollution/

[12] S. Berg, "Types of pollutants,"

https://sciencing.com/types-pollutants-5270696.html

[13] "Environmental problems,"

https://www.conserve-energy-future.com/15-current-environmental-problems.php

[14] "Air pollution: Current and future challenges," Unknown Source

[15] E. Konduracka, "A link between environmental pollution and civilization disorders: A mini review," *Reviews on Environmental Health,* vol. 34, no. 3, May 2019.

[16] "Pollution prevention (P2)," https://www.epa.gov/p2

[17] M. N. O. Sadiku, N. K. Ampah, and S. M. Musa, "Green waste disposal," *International Journal of Trend in Scientific Research and Development*, vol. 3, no. 2, January-February, 2019.

[18] A. W. Murch, "Public concern for environmental pollution," *The Public Opinion Quarterly*, vol. 35, no. 1, Spring 1971, pp. 100-106.

[19] S. Mueller, "Green technology and its effect on the modern world," *Bachelor's Thesis*, Oulu University of Applied Sciences, Spring 2017.

[20] Rinkesh, "What is environmental pollution?"

https://www.conserve-energy-future.com/causes-and-effects-of-environmental-pollution.php

[21] D. Briggs, J. J. Abellan, and D.Fecht, "Environmental inequity in England: Small area associations between socio-economic status and environmental pollution,"

Social Science & Medicine, vol. 67, no. 10, November 2008, pp. 1612-1629.

[22] N. Gayner, "Environmental pollution settlements: A London perspective,"

Environmental Claims Journal, vol. 11, no. 3, 1999, pp. 3-16.

[23] H. Yuanjun and Z. Zhongxing, "Environmental pollution and control measures in China," *Ambio*, vol. 16, no. 5, 1987, pp. 257-261

[24] Y. Feng et al., "Environmental pollution liability insurance in China: In need of strong government backing," *Ambio*, vol. 43, no. 5, September 2014, pp. 687-702.

[25] "Study Session 7 pollution: Types, sources and characteristics,"

https://www.open.edu/openlearncreate/mod/oucontent/view.php?id=79946&printable=1

[26] S. TuranKatircioglu, " International tourism, energy consumption, and environmental pollution: The case of Turkey," *Renewable and Sustainable Energy Reviews,* vol. 36, August 2014, pp. 180-187.

[27] R. R. Appannagari, "Environmental pollution causes and consequences: A study,"

North Asian International Research Journal of Social Science & Humanities, vol. 3, no. 8, August 2017, pp. 151-161.

[28] A. Turkman and M. Goto, "Present status of environmental pollution in Japan," *Industrial & Environmental Crisis Quarterly*, vol. 8, no. 2, 1994, pp. 129- 139.

[29] B. Heppell and R. Wiltshire, "Environmental Pollution and Japan," *Teaching Geography*, vol. 15, no. 2, April 1990, pp. 67-70.

[30] A. O. Ikpah, "Oil and gas industry and environmental pollution application of systems reliability analysis for the evaluation of the status of environmental pollution control in the Nigerian petroleum industry," *Doctoral Dissertation,* The University of Texas at Dallas, May 1981.

[31] B. S. Miapyen and U. Bozkurt, "Capital, the state, and environmental pollution in Nigeria," *Sage Open,* October-December 2020, pp. 1–10.

[32] I. Yuksel and A. Dorum, "The role of hydropower in energy utilization

and environmental pollution in Turkey," *Energy Sources, Part A: Recovery, Utilization, and Environmental Effect*s, vol. 33, no. 13, 2011, pp.1221-1229.

[33] "Top environmental pollutants in Canada,"

https://www.weconserve.ca/2019/03/20/top-environmental-pollutants-in-canada/

[34] S. A. Oke, "On the environmental pollution problem: A review," *Journal of Environmental Engineering and Landscape Management*, vol. 12, no. 3, 2004, pp. 108-113.

[35] H. S. Rathore and L. M. L. Nollet (eds.), *Pesticides: Evaluation of Environmental Pollution.* Boca Raton, FL: CRC Press, 2012.

[36] P. Singh, A. Kumar, and A. Borthakur (eds.), *Abatement of Environmental Pollutants: Trends and Strategies.* Elsevier, 2019.

[37] P. R. Trivedi, *Environmental Pollution and Control.* Ashish Publishing House, 2004.

[38] F. R. Spellman, *The Science of Environmental Pollution.* Boca Raton, FL: CRC Press, 3rd edition, 2017.

[39] A. Vesilind and T. D. DiStefano, *Controlling Environmental Pollution: An Introduction to the Technologies, History, and Ethics.* DEStech Publications, 2006.

[40] A. Chauhan, *Environmental Pollution and Management.* Delhi, India:

I.K. International Publishing House, 2019.

[41] G. Best, *Environmental Pollution Studies.* Liverpool University Press, 2000.

[42] R. E. McKinney, *Environmental Pollution Control Microbiology: A Fifty-Year Perspective.* Boca Raton, FL: CRC Press, 2004.

[43] J. G. Speth, *Environmental Pollution: A Long-Term Perspective.* Washington, D C: National Geographic Society, 1988.

[44] R. O. Gilbert, *Statistical Methods for Environmental Pollution Monitoring.* New York: John Wiley & Sons, 1987.

[45] M. K. Hill, *Understanding Environmental Pollution.* Cambridge, UK: Cambridge University Press, 3rd edition, 2020.

[46] B. Alloway and D. C. Ayres, *Chemical Principles of Environmental Pollution.* London, UK: Blackie Academic & Professional, 2nd edition, 1997.

[47] S. M. Khopkar, *Environmental Pollution Monitoring and Control.* New Delhi, India: New Age International Limited Publishers, 2007.

[48] J. Kaiser, J. E. Klanning, and L. E. Erickson, *Bioindicators and Biomarkers of Environmental Pollution and Risk Assessment.* Science Publishers, 2001.

[49] P. A. Vesilind, J. J. Peirce, and R. F. Weiner, *Environmental Pollution and Control.* Boston, MA: Butterworth-Heinemann, 4th edition, 1998.

[50] C. S. Rao, *Environmental Pollution Control Engineering.* New York: John Wiley & Sons, 1991.

[51] A. Farmer, Managing Environmental Pollution. London: Routledge, 1997.

[52] The National Research Council, *Global Sources of Local Pollution: An*

Assessment of Long-Range Transport of Key Air Pollutants to and from the United

States. National Academies Press. 2010.

[53] J. N. Bhakta, S. Lahiri and B. B. Jana (eds.), *Green Technology for Bioremediation of Environmental Pollution.* Nova Science Publishers, 2019.

[54] C. S. Rao, *Environmental Pollution Control Engineering.* New Delhi, India: New Age International Limited Publishers, 2nd edition, 2007.

[55] R. Tykva and D. Berg (eds.), *Man-Made and Natural Radioactivity in Environmental Pollution and Radiochronology.* Springer, 2004.

[56] C. A. Edwards (ed.), *Environmental Pollution by Pesticides.* Springer, 1973.

[57] J. Rieuwerts, *The Elements of Environmental Pollution.* London: Routledge, 2015.

[58] M.W. Holdgate, *A Perspective of Environmental Pollution.* Cambridge University Press, 1980.

[59] J. J. Peirce, P. A. Vesilind, R. Weiner, *Environmental Pollution and Control.* Elsevier, 4th edition, 1998.

[60] H. S. Rathore and L. M. L. Nollet, *Pesticides: Evaluation of Environmental Pollution.* Boca Raton, FL: CRC Press, 2012.

[62] M. W. Holdgate, *A Perspective of Environmental Pollution.* Cambridge, UK: Cambridge University Press, 1980.

CHAPTER 7

ENVIROMENTAL HEALTH

"Saving our planet, lifting people out of poverty, advancing economic growth... these are one and the same fight. We must connect the dots between climate change, water scarcity, energy shortages, global health, food security and women's empowerment. Solutions to one problem must be solutions for all." – Ban Ki-moon

7.1 INTRODUCTION

Humans constantly interact with the environment. Our health largely depends on the environment around us. The term "environment" is used broadly to include everything external to ourselves, including the physical, natural, social, and behavioral environments. It includes the air we breathe, the water we drink, the food we eat, and the places where we live, work, school, and play [1]. There are two kinds of environment: natural environment and built environment, which comprises the areas that are influenced by humans. Many aspects of our environment (both built and natural environment) can impact on our health. Besides the two types of environment mentioned earlier, we can also think of the physical environment (e.g., air quality), the social environment (e.g., social capital), the economic environment (e.g., unemployment rates), the cultural environment (e.g., role of women), or the political environment (e.g., access to opportunities and institutions). These environments have been shown to influence human health. Humans and environment interact in a two-way reciprocity: human activities affect environmental quality, while environmental conditions affect human health [2].

Environmental health (EH) is the discipline that is concerned with the natural and built environments for the benefit of human health. It addresses all the physical, chemical, and biological factors external to a person and impacting behaviors. It is a key component of any comprehensive public health system. The discipline promotes policies and programs to reduce the release of harmful substances in air, water, soil, and food in order to protect people and provide healthy environments [3].

This chapter provides an introduction to environmental health. It begins by discussing what environmental health is all about, and why it is important. It describes the duties of environmental health professionals. It addresses global environmental health. It highlights some of the benefits and challenges of EH. The last section concludes with comments.

7.2 WHAT IS ENVIRONMENTAL HEALTH?

There is a connection between the environment and the health of individuals living in the environment. Environmental factors have been identified as the root cause of a significant burden of death, disease and disability. Figure 7.1 shows some major causes of environmental health problems [4].

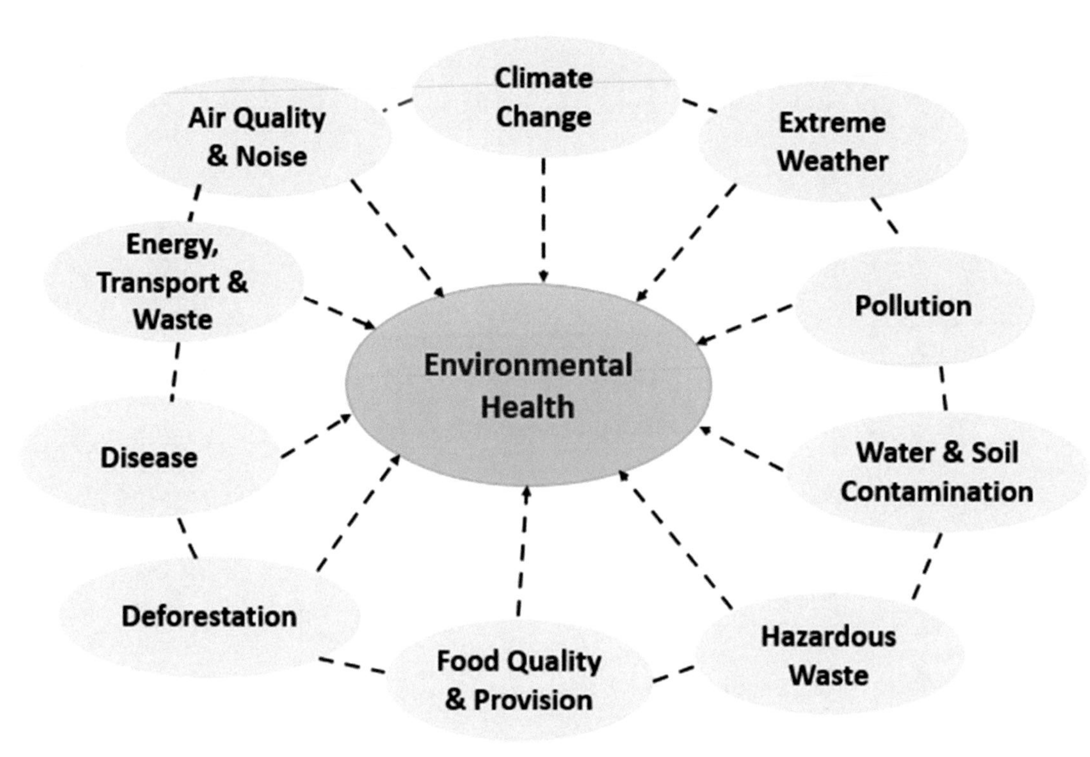

Figure 7.1 Major causes of environmental health problems [4].

Environmental health (EH) refers to the prevention of disease and injury related to the interactions between people and their environment. It comprises those aspects of human health (mental, physical, and social well-being) that are determined by physical, chemical, biological, and psychosocial factors in the natural and built environment. Addresses all the physical, chemical, and biological factors external to a person, and all the related factors impacting behaviors [5]. It is the branch of public health that focuses on all aspects of the natural and built environment (the areas that are influenced by humans, as opposed to natural environment) affecting human health and disease. It may also be regarded to the theory and practice of assessing and controlling factors in the environment that can potentially affect directly or indirectly human health or cause disease.

Environmental health is a broad and complex field which necessitates concerted multidisciplinary approaches to understanding and addressing environmentally influenced health outcomes. The five main disciplines that contribute to EH are environmental epidemiology, toxicology, exposure science, environmental engineering, and environmental law.

A lot of adults and children die due to preventable environmental factors, which include exposure to hazardous substances in the air, water, soil, food, climate change, occupational hazards, and built environment. Creating healthy environments requires understanding the effects of exposure to environmental hazards on people's health. Environmental hazards (like water and air pollution, extreme weather, or chemical exposures) can affect human health in a number of ways. The effect is influenced by many factors such as dose, duration, exposure route, and personal traits. After exposure to an environmental hazard, it may be possible to determine how much of a substance has got into a person's body. The exposed person may or may not become ill depending on the amount of penetration. Figure 7.2 shows how chemicals or substances in the environment can get into the human body [6]. Environmental health professionals often use the term "exposure" to describe the total amount of substance that comes in direct contact with your body.

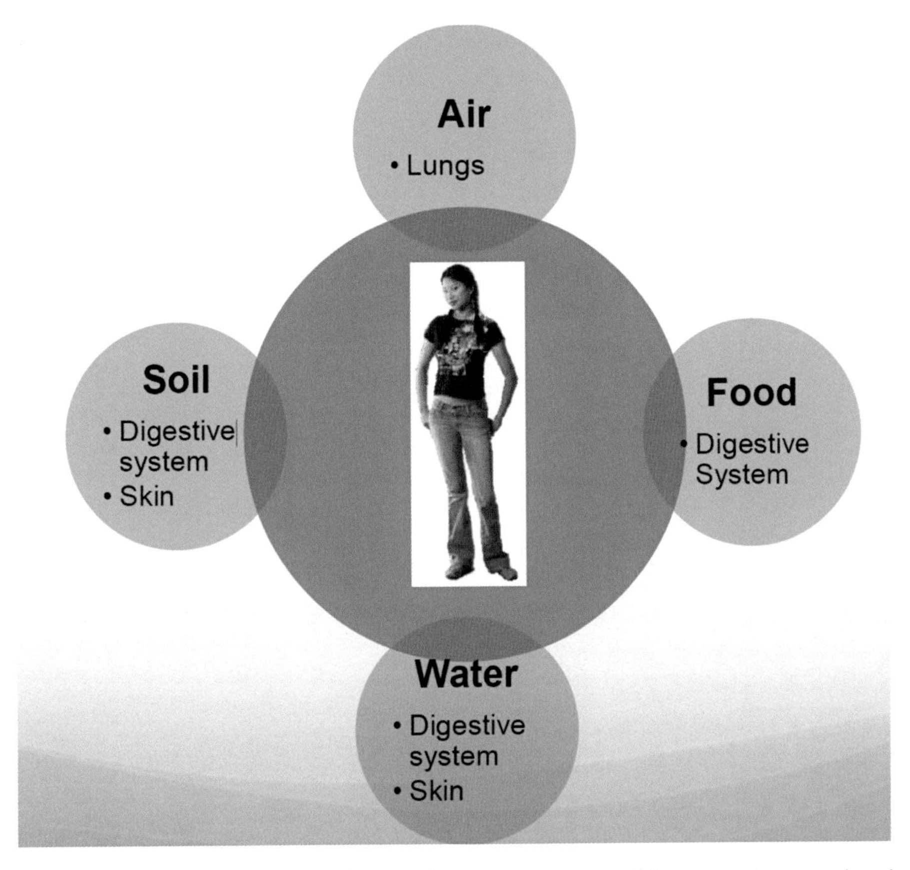

Figure 7.2 How substances in the environment can get into the human body [6].

Environmental health services are the services which implement environmental health policies through monitoring and control activities. These are not a dispensable luxury but an indispensable necessity. They are important in providing basic needs like safe drinking water, clean air, chemical and food safety, solid waste management, radiation protection, healthy and affordable housing, etc.

7.3 WHY IS ENVIRONMENTAL HEALTH IMPORTANT?

Environmental health (EH) focuses on the interrelationships between people and their environment, promotes a safe and healthy environment. Healthy environments could prevent almost one quarter of the global burden of disease. Thus, maintaining a healthy environment is central to increasing quality of life and years of healthy life. EH addresses the societal and environmental factors that increase the likelihood of exposure and disease.

Specifically, the issues (some of which are emerging) that environmental health addresses include [7]:

- *Air quality:* This include both ambient outdoor air and indoor air quality, which also comprises concerns about environmental tobacco smoke. Decreasing air pollution is important in creating a healthy environment. Tracking air pollution can help people understand how often they are exposed to unhealthy levels of air pollution.
- *Clean Drinking Water:* Drinking water quality is an important because contamination in a single system can expose many people simultaneously. Protecting water sources and minimizing exposure to contaminated water are important in environmental health. Diseases can be minimized by improving water quality and increasing access to pure water and sanitation facilities.

- *Household Energy Supply:* This is a major environmental health issue because of the harmful effects of biomass and coal smoke. Alternative fuels for cooking and heating need to be made available and electricity may be used instead.
- *Climate Change*: This constitutes a large-scale and highly inequitable environmental risk. It is an emerging threat to global public health. It challenges environmental health and the sustainability of global development. It affects sea level, infectious disease, air quality, and natural disasters such as floods, droughts, and storms. It is hard to link climate change to a specific health issue. Climate change threatens our life as shown in Figure 7.3 [8].

Figure 7.3 Climate change threatens our life [8].

- *The Built Environment*: This refers to the areas that are influenced by humans, as distinct from natural environment. It affects behaviors, physical activity patterns, social networks, and access to resources. A lot of people spend most of their time at home, work, or school. Maintaining healthy communities is crucial to environmental health.
- *Environmental Racism:* Environmental racism refers to the institutional rules, regulations, policies or government and/or corporate decisions that deliberately target certain communities for locally undesirable land uses and lax enforcement of zoning and environmental laws, resulting in communities being disproportionately exposed to toxic and hazardous waste based upon race [59]. Environmental racism may be caused by an intentional neglect of the government on people of color. For instance, it is a well-documented fact that communities of color and low-income communities are disproportionately impacted by polluting industries (and very specifically, hazardous waste facilities). This disproportionately affects different groups globally, especially the most marginalized groups of any nation.
- *Nanotechnology*: Technically savvy professionals realize that nanotechnology is having a major impact on science, technology, and society. The potential impact of nanotechnology offers

possible improvements on clean energy (reducing greenhouse gases and hazardous wastes), manufacturing, environmental risk assessment, disease prevention, detection, and treatment. Carbon nanotubes are regarded as one of the most promising materials in nanotechnology and has many technological applications.

7.4 ENVIRONMENTAL HEALTH PROFESSIONALS

They are also known as environmental health officers, public health inspectors, environmental health specialists or environmental health practitioners. In some countries, physicians and veterinarians are involved in environmental health. EH is a broad discipline where professionals perform many tasks such as research, investigations, outreach and education, environmental cleanups, etc. EH professionals work in both public and private sectors in organization, management, education, enforcement, consultation, and emergency response. Figure 7.4 shows some environmental health workers [9].

Figure 7.4 Environmental health workers [9].

The National Environmental Health Association (NEHA) represents more than 5,000 environmental health professionals in the United States and around the world. NEHA believes that health is the basis of every community's prosperity and that providing safe food, safe drinking water, clean air, safe sewage disposal, emergency response, and healthy living and workplace environments are basic necessities for communities. The goal of NEHA is to provide current and relevant information on environmental health and to partner on innovative initiatives [10].

Environmental health professionals are needed in different areas. In US, for example, some EH professionals work for agencies or centers that are primarily interested in environmental health issues. These include Environmental Protection Agency, Federal Emergency Management Agency, National Cancer Institute, Centers for Disease Control and Prevention, American Association of Poison Control Centers, National Center for Environmental Health, and National Oceanic & Atmospheric Administration, Food Drug Administration, and the Department of Health Services.

The task of EH professionals is to monitor air quality, water and noise pollution, toxic substances and pesticides, conduct restaurant inspections, and promote healthy land use and housing. They also prevent environmental health hazards and the promotion of public health and the environment in the following areas: food protection, housing, institutional environmental health, land use, recreational swimming areas and waters, electromagnetic radiation control, hazardous materials management, underground storage tank control, drinking water quality, water sanitation, emergency preparedness, and milk and dairy sanitation pursuant [11]. In order words, environmental health practices include [1,12]:

- Monitoring data on environmental hazards and health effects
- Identifying environmental hazards
- Investigating environmental concerns
- Stopping or lessening hazards
- Assessing individuals' exposure to hazards
- Researching possible health effects related to exposures
- Preventing and/or lessening health effects
- Diagnosing and treating health effects
- Investigating, sampling, measuring, and assessing hazardous environmental agents in various environmental media and settings
- Recommending and applying protective interventions that control hazards to health
- Developing, promoting, and enforcing guidelines, policies, laws, and regulations
- Developing and providing health communications and educational materials
- Managing and leading environmental health units within organizations
- Performing systems analysis
- Engaging community members to understand, address, and resolve problems
- Reviewing construction and land use plans, and make recommendations
- Interpreting research, utilizing science, and evidence to understand the relationship between health and environment
- Interpreting data and preparing technical summaries and reports

7.5 ENVIRONMENTAL HEALTH INTERVENTIONS

Environmental health is a diverse broad of professional practice that protects us from the many environmental hazards. It is targeted towards preventing diseases and creating health supportive environments. Food safety, clean water, air quality, and hazardous waste have been the traditional domains of environmental public health. Environmental health professionals are in the front line of public health by preventing disease and injury. They use the following tools in treating or solving environmental health problems. It is needless to say that each of these intervention tools has its own limitations.

- *Environmental Health Promotion:* Health promotion is crucial to environmental health practice. This lies at the intersection between the disciplines and can be regarded as any planned process employing comprehensive promotion approaches to assess, correct, control, and prevent environmental problems. Health promotion intervention strategies should target a range of outcomes and occur at multiple levels using a social ecological framework.
- *Environmental Health Engineering*: This is the application of engineering principles to reduce the disease burden by reducing environmental risks to health. Traditionally, EHE focuses on domestic water supply and excreta management plus other engineering utility and infrastructure services such as stormwater drainage, solid waste management, and indoor environmental quality. The potential utility of EHE is underscored by global health status statistics. The World Health Organization (WHO) estimates that 24% of the entire global disease burden and 23% of all deaths are attributable to environmental factors. Two key EHE considerations that are

often underrated or neglected in connection with the planning and provision of EHE services, especially in developing countries, are (1) the diversity of EHE challenges and service needs and (2) the evolutionary nature of most of the EHE challenges and service needs [2].

- *Environmental Health Tracking:* This is an emerging area that spans the traditional sectors of public health and environmental protection. Its goal is to provide more information about how the environment affects health. It incorporates elements of environmental monitoring, community health, and public health surveillance of diseases. Environmental health tracking incorporates elements of environmental monitoring, public health surveillance, and community health. It incorporates public health surveillance of diseases that are caused by environmental factors. It incorporates elements of community health as it is intended to provide information relevant to communities seeking to address environmental health issues [2]. Environmental monitoring includes collection of environmental media (air, water, soil) using devices that detect exposures to hazardous agents. Coordinated surveillance systems can assist public health efforts to prevent and control disease, injury, and disability.
- *Risk Assessment:* This is fundamental to traditional environmental health practice. Risk assessments are often used to ascertain the presence of toxic substances and to evaluate whether these toxic substances pose a risk to the environment and human health. There are four basic steps in risk assessment: (1) hazard identification, (2) dose-response assessment, (3) exposure assessment, and (4) risk characterization. Once a significant risk is determined to be present, some risk management strategies may be employed. Four major types of strategies are: (1) control of the source, (2) control along the path, (3) control at the level of the person, (4) secondary prevention. For example, these strategies may include posting warning signs about fishing, swimming, and picnicking near contaminated rivers [13].

7.6 ENVIRONMENTAL HEALTH IMPACT ASSESSMENTS

It is well known that exposure to environmental factors can cause adverse health effects or even death. Environmental effects on health have always been regarded as multi-facetted. Environmental health effects vary significantly with regard to their severity, type of disease, and duration. Many environmental health professionals are concerned about the impact of environmental stressors on health. Quantifying or assessing the impacts of environmental degradation on human health is important for the development of well-informed policies by the health sector. Integrated environmental health impact assessment (IEHIA) aims to support policy making by comprehensively assessing environmental health effects [14]. Although traditional environmental health impact assessments often have to deal with substantial uncertainties, they have provided good service in support of policy development, standard setting, and regulation of hazardous chemicals or practices. Outcomes of the assessment is often presented as measures of impact. For example, the toxicity scale in Figure 7.5 measures how dangerous is the substance in the environment [15].

Toxicity Rating	Signal Words on Package	Symbol on Package
Highly Toxic	**DANGER** or **POISON**	
Moderately Toxic	**WARNING**	
Slightly Toxic	**CAUTION**	
Not Toxic	none	

Figure 7.5 Toxicity scale [15].

The environmental burden of disease (EBD) is a measure of environmental health used in Europe. EBD assessment is aimed to support efficient policy development and resource allocation. It is also helpful for setting priorities in environmental health policies and risk management [16].

7.7 GLOBAL ENVIRONMENTAL HEALTH

Environmental health is a broad and complex problem as it has a lot of magnitude linked nationally, regionally, and globally. Surveillance of diseases and monitoring of environmental hazards are no longer the responsibility of each nation as they are all linked together by one environment. EH is a global issue that requires every nation to work together on regional, national, and global policies for dealing with related EH issues. Since the developed countries own more than 90% of the world wealth, they must be at the forefront of any action [2]. Figure 7.6 shows a typical example of global environmental health problem [17].

Figure 7.6 Global environmental health problem [17].

Global change drivers include climate change, population growth, aging population, urbanization, globalization, consumerism, improvements in living standard, industrialization, and energy intensive lifestyle [18]. Urbanization has led to a shift in consumption from wild game to industrially reared beef, pork, and chickens. It limits possibilities for household food production and contributes to environmental damage through long distance food transportation. It has also caused workplaces to become health hazards because of toxic products, injury and ergonomic hazards, noise, external pollution, and traffic generation [19]. Several millions of people around the world die yearly because they live or work in unhealthy environments. Increased mortality and morbidity of humans due to exposure to chemicals are recorded especially in developing countries.

Global effort is being made to reduce people's exposure to harmful pollutants in air, water, soil, food, and materials in school, homes, and workplaces. In 2002, the World Health Organization (WHO) and the United Nations Environment Programme (UNEP) joined forces to launch the Health and Environment Linkages Initiative (HELI). Also, the WHO and the National Institute of Environmental Health Sciences (NIEHS) have worked together for more than 30 years to enhance global environmental health through research, training, and information-sharing activities. The WHO reckons that it is the home, not the clinic, that is the key to a better healthcare delivery system. No country in the world could ever overcome the challenges and problems of environmental health until its economic, social, and environmental activities gravitate towards a health-conscious economy. We now present how some nations handle environmental health.

- *United States:* Many states in the US require that environmental health practitioners have a bachelor's degree and professional licenses in order to practice environmental health. Hundreds of new chemicals are introduced to the US market every year. Some of these chemicals may present unexpected challenges to human health. With global chemical production projected to double over the next 24 years and an antiquated US chemicals policy, federal policies that shape the priorities of the US chemical enterprise will be a cornerstone of sustainability [20].

The United States will face growing health, environmental, and economic problems related to chemical exposures and pollution. Blood lead exposure levels can affect a child's academic performance. US government agencies include [1]:

- Census Bureau
- Department of Education
- Environmental Protection Agency
- Federal Emergency Management Agency
- National Aeronautics & Space Administration
- National Cancer Institute
- National Center for Education Statistics
- National Oceanic & Atmospheric Administration

- *United Kingdom:* The environmental health profession had its modern-day roots in the sanitary and public health movement of the United Kingdom. In the UK, environmental health practitioners are required to have a graduate degree in environmental health and be certified and registered with the Chartered Institute of Environmental Health or the Royal Environmental Health Institute of Scotland.
- *Malaysia:* EH issues experienced by Malaysia originate mainly from atmospheric pollution, water pollution, climate change, ozone depletion, solid waste management, and hazardous waste management. Urban air pollution and trans-boundary haze problems are causing chronic health problems. Malaysia is also having environmental problems with its water resources and manufacturing. The rural areas are reported to have a clean environment. Malaysia's ability to deal with its environmental health issues is remarkable. The country continues to play an active role in planning and implementing environmental health activities at a national level [2].
- *China:* Environmental health crisis is testing the resilience of China. Poised to become a superpower, China confronts some serious environmental threats. Water in China poses a triple threat: supply is scarce in the populous north, flooding endangers lives in the south, and growing industrial pollution jeopardizes the entire nation [21]. Air and water pollution is a major source of morbidity and mortality in China. Air quality in China's cities is among the worst in the world. Biomass fuel and coal are burned for cooking and heating in almost all households, resulting in air pollution. Industrialization in China is associated with increase in energy use and industrial waste, with severe effects on health. Global climate change will intensify China's environmental health problems. Facing the emerging environmental dilemmas, China has committed substantial resources to improve both local and global environments [22].
- *Japan:* Mercury poisoning occurred among Japanese fishing populations. This along with pollution from the paper industry caused environmental accumulation of methylmercury in food chains. Today, thousands of lakes and rivers worldwide are polluted with methylmercury. Seafood from Japan was found to contain an average methylmercury concentration. Mercury analyses of fish begun in Japan in the 1960s. The Environment Agency issued a notice on certification of patients in 1977. In Minamata, Japan, a series of infants who suffered congenital mercury poisoning in their mother's womb was recorded by the investigative team from Kumamoto University. The criteria for Minamata disease diagnosis became the object of discussion and legal proceedings [23].
- *Canada:* In Canada, environmental health practitioners must have an approved bachelor's degree in environmental health along with the national professional certificate, the Certificate in Public Health Inspection. Recent data from Canada indicates that high-use drugs include

acetominophen, acetylsalicylic acid, ibuprofen, naproxen, and carbamazepine. Large amounts of veterinary medicines, such as antibacterials, antifungals and parasiticides from aquaculture and agriculture, may also contribute to the stress on the environment. Once released into the environment, pharmaceuticals will be distributed to air, water, and soil. Degradation varies and depends on chemistry, biology, and climatic conditions. The entire population does not react in the same manner when exposed to environmental contaminants.

- *Europe:* Since the 1970s, integration and precaution have been underpinning ideals of the Environment Action Plan of the European Union. In 2003, the World Health Organization established national environmental health action plans, which were adopted into a European plan. The European Environment Health Action Plan nevertheless marks an important turning point in health policy as it highlights the need for a better, more timely, and more integrated environment health policy. It emphasizes also the need for developing and applying new, integrated methods for assessment to support environmental health policies [24]. Integrated measures of population health, such as environmental burden of disease (EBD), are useful for setting priorities in environmental health policies. The Environmental Burden of Disease in European countries (EBoDE) project aimed to provide harmonized EBD assessments for participating countries. One of its aims is to assess data availability and method of applicability for this type of EBD assessment [15].
- *Nigeria:* Roughly 25% to 30% of Nigerians, mainly top government officials and other rich and privileged people, enjoy a decent quality of urban life. Most of the laws and regulations for environmental health appear to be reminders of colonial oppression and have very little current relevance today. The control of all land was vested in the government. The main source of air pollution is exposure to toxic fumes from cooking fires and stoves inside poorly ventilated homes [25].

7.9 BENEFITS AND CHALLENGES

The benefits of environmental health interventions are hard to measure because it requires an understanding of the relationship between the environmental exposure due to pollutants and health outcomes, which is often uncertain. Protecting the environment implies that the air safe to breathe, water is pure to drink, and land is free of toxins. Environmental health should play a major role in educating our community about preparedness, especially before any emergency such as a pandemic. The World Health Organization has created a task force in response to growing concerns about the children's environmental health internationally. Other benefits of environmental health and protection include [26]:

- enhanced economic status
- enhanced productivity
- enhanced educational achievement
- less social problems
- a more livable environment
- better quality of life
- reduced disease and disability
- reduced health care costs

Although much progress has been made in protecting the environment, we are yet to overcome some challenges. Environmental health is a broad, complex, and multi-facetted field that addresses all aspects of the environment that can affect human health. There are gaps in information about how the environment affects human health. Some health effects are well known, while others are suspected.

Linking climate change to a specific health problem is difficult. Maintaining environmental quality is a difficult task. The growing complexity of environmental issues facing policy-makers has highlighted the need for more integrated methods of assessment to guide decision-taking. Currently, most nations do not have surveillance system for heart attack or coronary heart disease. Children, women, the elders, ethnic minority populations, indigenous peoples, and the poor are vulnerable to the development of environmentally linked disease, disability, or diminished quality of life. None of these challenges are insurmountable.

7.9 CONCLUSION

Environmental health is the interconnections between people and their environment. It is mainly concerned with the biological, chemical, and physical influences on human health. It addresses all the physical, chemical, and biological factors impacting behaviors. It is a dynamic and evolving field. Reducing exposure to toxic substances and hazardous wastes is fundamental to environmental health. Legislations and policies to reduce different types of pollution can help prevent or minimize several health problems. Modern cities can improve health by providing safe drinking water, reducing air pollution from household cooking and heating, reducing traffic injury hazards, improving working environment, and reducing heat stress because of global climate change [20].

World Environmental Health Day is observed globally on 26th September every year. Since January 2019, the Department of Public Health Sciences has been offering training on environmental health. Future environmental health problems will require a new generation of well-educated and trained professionals. Some academic institutions offer master's degrees online on EH. For more information about environmental health, one should consult the books in [2,27-57] and the following related journals [58]:

- *Environmental Health*
- *Environmental Health Perspectives*
- *Environmental Determinants of Human*
- *Current Environmental Health Reports*
- *Archives of Environmental Health: An International Journal*
- *International Journal of Environmental Health Research*
- *International Journal of Environmental Research and Public Health*
- *International Journal of Hygiene and Environmental Health*
- *International Journal of Occupational and Environmental Health*
- *Journal of Environmental Health*
- *Journal of Occupational and Environmental Medicine*
- *Journal of Toxicology and Environmental Health, Part B*
- *Reviews on Environmental Health*

REFERENCES

[1] "Introduction to environmental public health tracking,"

https://www.cdc.gov/nceh/tracking/tracking-intro.html

[2] P. Smith, "Environmental health," in J. O. Nriagu (ed.),

Encyclopedia of Environmental Health. Elsevier, 5-Volumes, 2nd edition, 2019.

[3] M. N. O. Sadiku, Y. P. Akhare, A. Ajayi-Majebi, and S. M. Musa, "Environmental health: An introduction," *International Journal of Trend in Research and Development*, vol. 7, no. 6, Nov.-Dec.2020, pp. 66-70.

[4] "Background, aims & rationale – Environmental,"
https://sites.google.com/site/healthypolisnetwork/rationale

[5] P. K. Chopra and G. K. Kanji, "Environmental health: Assessing
risks to society," *Total Quality Management,* vol. 22, no. 4, 2011, pp. 461-489.

[6] What is environmental health?"
https://phpa.health.maryland.gov/OEHFP/EH/Shared%20Documents/curriculum/8GRD_EH_PP.pdf

[7] "Environmental health,"
https://www.healthypeople.gov/2020/topics-objectives/topic/environmental-health

[8] "Chapter 1 Introduction: The environment at risk," Unknown Source.

[9] "Environmental consultants Calgary and Western Canada,"
https://www.pinterest.ca/pin/805299977101342639/

[10] "National Environmental Health Association," *Journal of Environmental Health,* vol. 77, no. 4, November 2014, p. 62

[11] "Environmental health," *Wikipedia*, the free encyclopedia
https://en.wikipedia.org/wiki/Environmental_health

[12] "Definitions of environmental health,"
https://www.neha.org/about-neha/definitions-environmental-health

[13] E. H. Howze, G. T. Baldwin, and M. C. Kegler, "Environmental health promotion: Bridging traditional environmental health and health promotion,"
https://pubmed.ncbi.nlm.nih.gov/15296627/

[14] A. B. Knol et al., "The use of expert elicitation in environmental health impact assessment: A seven step procedure," *Environmental Health,* vol. 9, Article number: 19, 2010.

[15] "What is environmental health? A student introduction,"
https://depts.washington.edu/ceeh/downloads/Intro_to_EH_slideset.pdf

[16] O. Hänninen et al., "Environmental burden of disease in Europe: Asessing nine risk factors in six countries," *Environmental Health Perspectives,* vol. 122, no. 5, 2014, pp. 439–446.

[17] https://ehe.jhu.edu/research/research-areas/

[18] M. A. Hanjra et al., "Wastewater irrigation and environmental health: Implications for water governance and public policy," *International Journal of Hygiene and Environmental Health, vol. 125, no. 3,* April 2012, pp. 255-269.

[19] T. Kjellstrom et al., "Urban environmental health hazards and health equity," *Journal of Urban Health,* vol. 84, 2007, pp. 86–97.

[20] M. P. Wilson and M. R. Schwarzman, "Toward a new U.S. chemicals policy: Rebuilding the foundation to advance new science, green chemistry, and environmental health," vol. 117, no. 8, August 2009.

[21] C. Wu, "Water pollution and human health in China," *Environmental Health Perspectives,* vol. 107, no. 4, April 1999, pp. 251-256.

[22] J. Zhang et al., "Environmental health in China: Progress towards clean air and safe water," *The Lancet,* vol. 375, no. 9720, March–April 2010, pp. 1055-1056.

[23] P. Grandjean et al., "Adverse effects of methylmercury: Environmental health research implications," *Environmental Health Perspectives,* vol. 118, no. 8, August 2010.

[24] D. J. Briggs, " A framework for integrated environmental health impact assessment of systemic risks," *Environmental Health,* vol 7, Article number: 61, 2008.

[25] G. I. Nwaka, "The urban informal sector in Nigeria: Towards economic development, environmental health, and social harmony," *Global Urban Development,* vol. 1, no.1, May 2005.

[26] L. Gordon, "Principles of environmental health administration,"

http://www.sanitarians.org/Resources/Documents/Morgan_Revision.pdf

[27] H. Koren, *Best Practices for Environmental Health: Environmental Pollution, Protection, Quality and Sustainability.* Routledge, 2017.

[28] L. L. Harrison *Environmental, Health, and Safety Auditing Handbook.* McGraw-Hill Professional, 2nd edition, 1995.

[29] S. Finn and L. R. O'Fallon (eds.), *Environmental Health Literacy.* Springer, 2019.

[30] D. L. Tsalev, *Atomic Absorption Spectrometry in Occupational and Environmental Health Practice.* Boca Raton, FL: CRC Press, 1995.

[31] J. B. Sullivan and G. R. Krieger, *Clinical Environmental Health and Toxic Exposures.* Linppcott Williams & Wilkins, 2001.

[32] B. Wisner and J. Adams (eds.), *Environmental Health in Emergencies And Disasters: A Practical Guide.* World Health Organization, 2002.

[33] D. Mackay and R. S. Boethling, *Handbook of Property Estimation Methods for Chemicals: Environmental Health Sciences.* Boca Raton, FL: CRC Press, 2000.

[34] H. P. Hynes and D. Brugge, *Community Research in Environmental Health: Studies in Science, Advocacy and Ethics.* Routledge, 2005.

[35] L. Theodore and R. R. Dupont, *Environmental Health and Hazard Risk Assessment: Principles and Calculations.* Boca Raton, FL: CRC Press, 2012.

[36] C. C. Sellers, *Hazards of the Job: From Industrial Disease to Environmental Health Science.* The University of North Carolina Press, 1997.

[37] N. I. Maxwell, *Understanding Environmental Health: How We Live in the World.* Burlington, MA: Jones and Bartlett Publishers, 2nd edition, 2014.

[38] E. Adams, J. Bartram, and Y. Chartier (eds.), *Essential Environmental Health Standards in Health Care.* World Health Organization, 2008.

[39] H. Koren and M. Bisesi, *Handbook of Environmental Health and*

Safety, Principles and Practices. Boca Raton, FL: CRC Press, volume 1, 4th ed., 2002.

[40] I. F. Goldstein and M. Goldstein, *How Much Risk? A Guide to Understanding Environmental Health Hazards.* New York, NY: Oxford University Press, 2002.

[41] R. J. Heinsohn and J. M. Cimbala, *Indoor Air Quality Engineering: Environmental Health and Control of Indoor Pollutants.* Boca Raton, FL: CRC Press, 2013.

[43] B. Hall and M. L. Kerr, *1991-1992 Green Index: A State-by-State Guide to the Nation's Environmental Health.* Island Press,1991.

[44] B. S. Levy et al. (eds.), *Occupational and Environmental Health: Recognizing and Preventing Disease and Injury.* Oxford University Press, 6th edition, 2011.

[45] A. Yassi et al., *Basic Environmental Health.* Oxford University Press, 2001.

[46] S. Cairncross and R. Feachem, *Environmental Health Engineering in the Tropics: An Introductory Text.* Wiley, 1983.

[47] H. Frumkin, *Environmental Health: From Global to Local (Public Health/Environmental Health).* Jossey-Bass, 3rd edition, 2016.

[48] R. H. Friis, *Essentials of Environmental Health (Essential Public Health).* Jones & Bartlett Learning, 3rd edition, 2018.

[49] D. W. Moeller, *Environmental Health.* Harvard University Press; 4th edition, 2011.

[50] P. Brown and L. Gibbs, *Toxic Exposures: Contested Illnesses and the Environmental Health Movement.* Columbia University Press, 2007.

[51] N. I. Maxwell, *Understanding Environmental Health: How We Live in the World.* Jones & Bartlett Learning, 2nd edition, 2013.

[52] R. D. Bulalrd, G. S. Johnson, and A. O. Torres, *Environmental Health and Racial Equity in the United States: Building Environmentally Just, Sustainable, and Livable Communities.* Amer Public Health Association, 2011.

[54] W. H. Bassett, *Clay's Handbook of Environmental Health.* London, UK:Son Press, 19th edition, 2004.

[55] M. McCally, *Life Support: The Environment and Human Health.* The MIT Press, 2002.

[56] M. Kumar and R. R. Tiwari (eds.), *Recent Trends and Advances in Environmental Health.* Nova Science Publishers, 2019.

[57] P. Purdom (ed.), *Environmental Health.* Academic Press, 2nd edition, 2013.

[58] "List of environmental journals," *Wikipedia,* the free encyclopedia

https://en.wikipedia.org/wiki/List_of_environmental_journals

[59] "Environmental Justice," http://greenaction.org/what-is-environmental-justice/

CHAPTER 8

ENVIRONMENTAL ECONOMICS

"A sustainable business is resource efficient, respects the environment and is a good neighbour." - Phil Harding

8.1 INTRODUCTION

All human societies have been shaped by their interaction with the environment. Human economic activities (production and consumption, buying and selling) have caused environmental irreparable damage and degradation. There is a high number of environmental crisis in many nations as a byproduct of economic activity. It is ironic that the nations with the highest levels of urban air pollution are also the fastest growing. High levels of environmental pollution imply immediate and direct health consequences for exposed populations [1].

Business is increasingly being held accountable for the social and environmental implications of its activities. Environmentalists, consumers, and policy-makers are increasingly demanding to account for resource used and to balance the proposed benefits with environmental sustainability. They are increasingly causing businesses to reconcile financial objectives with social responsibilities.

Economics is a science that deals with the allocation of scarce resources among competing uses. It is concerned about the satisfaction of man's unlimited wants with the scarce resources available. The environment includes all energy and material resources, aquatic ecosystems, and the atmosphere. It provides all life support systems with air, water, land, and the materials necessary to fulfill all developmental aspirations of man. Environmental economics attempts to bridge the traditional separation of economic and environmental considerations [2]. Figure 8.1 illustrates the relationship between environment and economy [3].

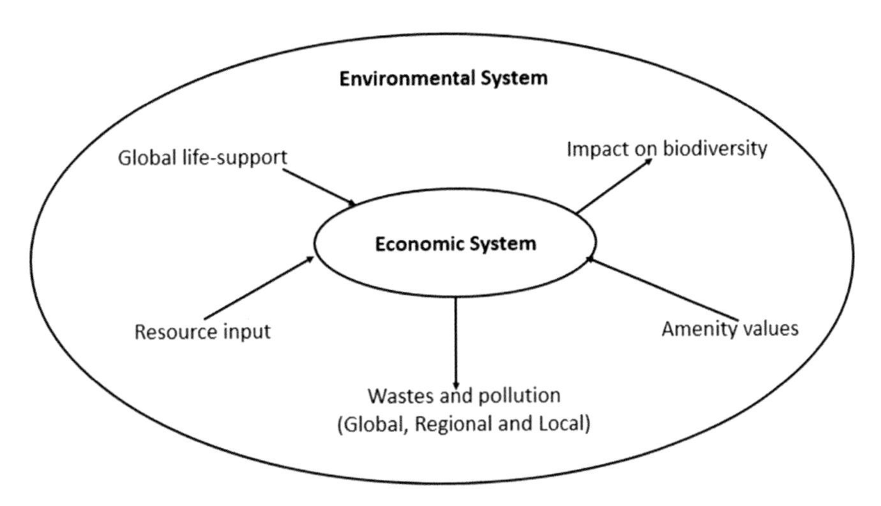

Figure 8.1 The environment- economy relationship [3].

This chapter provides an introduction to environmental economics. The chapter begins by explaining what environmental economics is all about. It discusses the scope of environmental economics. It covers the duties of environmental economists. It addresses global environmental economics. It highlights the benefits and challenges of environmental economics. The last section concludes with comments.

8.2 CONCEPT OF ENVIRONMENTAL ECONOMICS

Economics is concerned with making the best allocation of resources among competing alternatives. In addition to being a science of production and distribution, economics deals with decision making by agents, which include consumers, businesses, government agencies, and non-profit organizations. Mainstream economic theory assumed that economic systems were independent of environmental restraints and therefore these could be ignored. In the 1950s, a group of economists started to take the environment more seriously. When environmental economists speak of valuing the environment, they mean giving it a market price based on supply and demand and individual preferences. Resources that are not individually owned, such as the atmosphere, ocean, and areas of land are called public or social goods by economists [4]. Environmental economics, with its strong connection to public economics, is basically interested in the study of environmental public goods. It is currently characterized as an application of neoclassical economic theory to environmental problems. It may also be regarded as a synthesis of study of various branches of knowledge like science, economics, engineering, thermodynamics, philosophy, ethics, etc. The life of human beings is shaped by the living environment.

Environmental economics mainly deals with the relationship between the economy and the environment. It is a branch of economics that focuses on environmental problems and applies the tools of economics to address them. It studies the financial impact of environmental policies and the effects of environmental policies on the economy. It supports environmental policies to deal with air pollution, water quality, toxic substances, solid waste, and global warming. It considers issues such as the conservation and valuation of natural resources, pollution control, waste management, and recycling. Environmental economics can help us properly understand the value of the environment and take action to preserve and improve it [5].

The environmental revolution started in the late1960s when industrialization was experiencing a boom and pollution from industrial activities became an increasing concern. The growth of environmental economics in the 1970s was initially within the neoclassical paradigm, which was concerned with issues of market failure, inappropriate resource allocation and how to manage public goods. Since then, environmental economics has grown to be a major subdiscipline of economics.

Although environmental economics is related to ecological economics, they are different. Ecological economics focuses more explicitly on long-term environmental sustainability and issues [6]. It views the relationship of the economy and the environment as central. Ecological studies seeks harmony between nature and man, whereas economics focuses on the disharmony between man and nature. The relationship between the two is shown in Figure 8.2 [7].

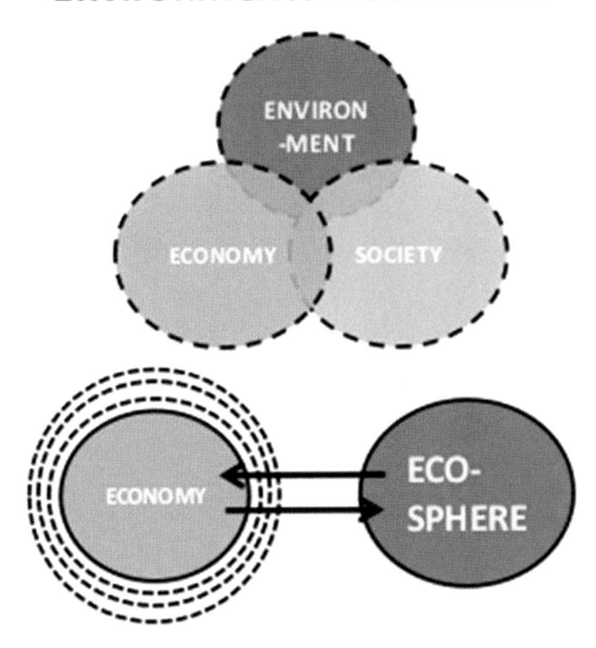

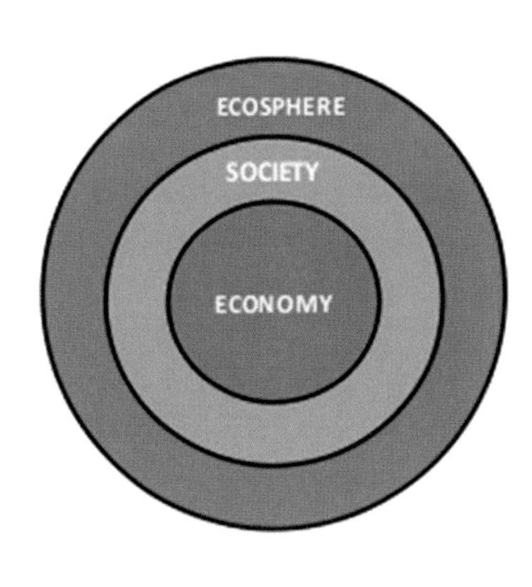

Figure 8.2 The relationship between environmental and ecological economics [7].

Environmental economics deals with the impact of economic activities on the environment [8]. It focuses on designing interventions that help attain economic efficiency when the market mechanism is not working properly or when market failure occurs. The advancement in technology continues to impose new demands on the environment. As shown in Figure 8.3, the fundamental principles of environmental economics are important in achieving sustainable development [9].

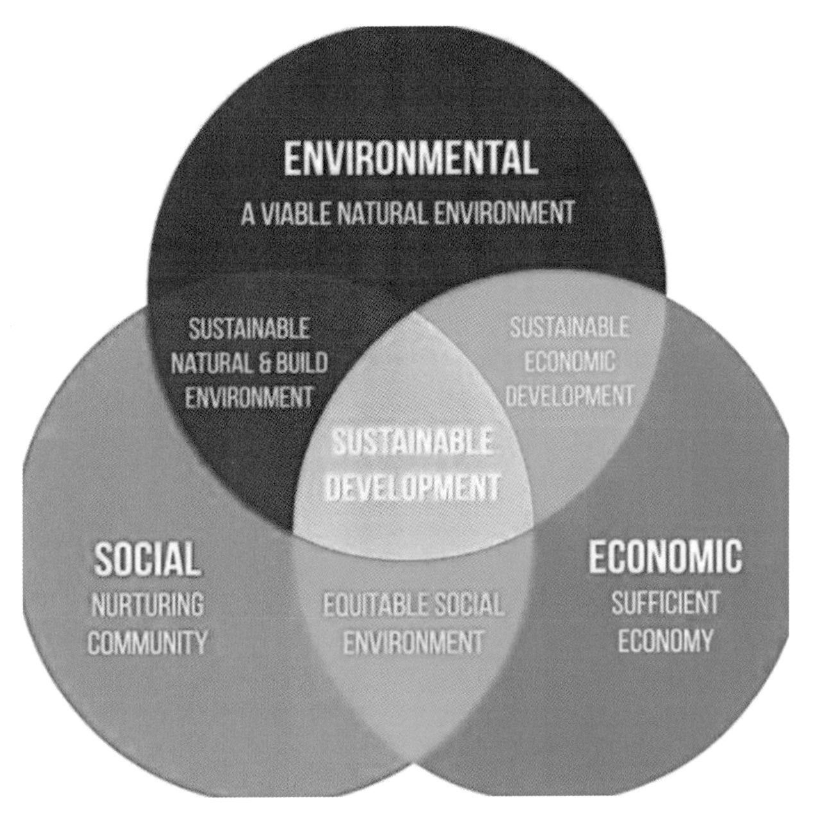

Figure 8.3 The relationship between environmental economics and sustainable development [9].

Environmental economics is a sub-discipline of economics that seeks to understand the economic causes of human impacts on the environment, such as atmospheric pollution. It involves the study of the decisions of agents (which include consumers, companies, government agencies, and non-profit organizations) that have environmental consequences and how to affect these decisions to achieve environmental objectives. This economic approach is used in dealing with such problems as environmental degradation, resource depletion, and global environmental change.

1.4 SCOPE OF ENVIRONMENTAL ECONOMICS

Environmental goods are aspects of the natural environment that hold value for individuals in society. The role of environmental economics in to study the impact of environmental policies and devise solutions to problems resulting from them. At the heart of environmental economics are a few concepts namely: opportunity costs, external costs, social costs, and private costs [10]. Thus, environmental economics embraces the following concepts [11].

1. *Sustainable Development:* This is regarded as the development that meets the needs of the present without compromising the ability of future generations to meet their own needs. The four basic components of sustainable development are economic growth, environmental protection, social equity, and institutional capacity. Environmental economists interpret sustainable development as development that maintains capital for future generations, where capital is the total of human capital (skills, knowledge, and technology) and human-made capital (such as buildings and machinery), as well as natural capital (environmental goods).
2. *Market Failure:* This occurs when the functioning of a perfect market is compromised. A market is essentially an exchange institution that serves society by organizing economic activity. A market failure occurs when the market does not allocate scarce resources to generate the greatest social welfare. For example, market failure happens when prices of timber, agriculture, and land do not provide an incentive to curtail habitat destruction.
3. *Externalities:* Externalities are at the center of environmental economics. These are unintentional effects on the environment resulting from economic activity. Inadvertent consequences of economic activity can either be negative or positive. Externalities are also another form of market failure. People who directly benefit from an economic resource without contributing to its establishment are known as "free riders."
4. *Valuation:* This aims at assigning dollar values to natural resources. This aspect of environmental economics helps to evaluate a variety of options in managing challenges with the use of environmental and natural resources. Value theory is the centerpiece and the foundation of economics. It must establish the economic conservation principle and provide a justification for it.
5. *Cost-Benefit Analysis:* This involves weighing the benefits of environmental regulations, i.e. CBA compares aggregate benefits with aggregate cost. Environmental economists incorporate environmental factors into project appraisals by modifying cost benefit analysis to include environmental costs and benefits. Hence, the best policy is one in which there is the greatest surplus of benefits over costs. Benefits include extra income, improved quality of life, clean water, and beaches, while costs include opportunity costs, internal and external costs, and externalities.

Environmental economists study and develop policy recommendations relating to these concepts.

8.3 ENVIRONMENTAL ECONOMISTS

Environmentalists are those who care for environment, realizing that any damage to the environment will affect the life of living things. It has been said that an economist is one who knows the price of everything and the value of nothing. Therefore, an environmental economist is one who studies and predicts the impact of environmental events on the local, national, and global economic scales. Environmental economists apply field research data to economic incentives and use the resulting models to assess labor, markets, trading, and outcomes. They are concerned with identifying specific problems to be rectified, but there can be many approaches and strategies to solve the same environmental problem. The strategies somewhat rely on state intervention in the market.

Environmental economists often receive the same training as economists. They see the environment as a form of natural capital that provides life support for humans. They apply the tools of economics to address environmental problems. They perform studies to determine the theoretical or empirical effects of environmental policies on the economy. They design appropriate environmental policies and analyze the effects and merits of existing or proposed policies. They research the economics of environmental issues such as renewable energy use, construction of new hydroelectric power plants, and pollution control measures. They explain resource use, measure the impact of resource degradation, and design policies to combat degradation.

They need to collaborate with professionals from other disciplines, since environmental economics is interdisciplinary in nature. Also because the nature of environmental goods often transcends national boundaries, environmental economics may require working with other professionals from other nations. Environmental economists may benefits from the following organizations: Association of Environmental and Resource Economists (AERE), American Economic Association (AEA), European Association for Environmental and Resource Economics (EAERE), International Society for Ecological Economics (ISEE), and Green Economics Institute.

Although the duties of an environmental economist may vary from job to job, the following list includes some typical duties [12]:

- Conduct research obtained from literature reviews, sample findings, and computer predictive data
- Analyze historical data and historical issues to formulate an economic theory to explain behavior, and apply to current circumstances
- Analyze ecological and economic trends and cycles, and use model data to relay information about future trends
- Look to change economic incentives so that people behave differently toward environmental concerns
- Consult with policymakers in regard to economic pressures that cause people and companies to conform to regulations
- Make predictions by identifying, collecting, analyzing, and using environmental and historical economic information
- Prepare reports and perform economic modelling
- Present reports to external stakeholders on economic performance and outcomes
- Consult with investors, policymakers, industry leaders, and other potential stakeholders on impacts of environmental decisions
- Assess and assign economic value to an industry or company's tasks, strategies, and past and future outcomes
- Provide consultation to agencies, professionals, or researchers based on the environmental economy perspective
- Establish workgroup culture for a positive and challenging work environment
- Mentor junior colleagues

8.4 GLOBAL ENVIRONMENTAL ECONOMICS

Environmental economics studies the economic effects of environmental policies around the world. It focuses on the design of environmental policies and its implementation. Environmental economist studies and predicts the impact of environmental events on the local, national, and global economic scopes. Mounting public concern with environmental degradation in the 1960s led governments in Western Europe, North America and Japan to address their national pollution problems. Enforcing agencies relied primarily on four types of instruments to curb industrial pollution: subsidies and soft loans, emissions standards, technology standards, and emissions charges. We now consider how environmental economics is practiced in some nations.

- *Unites States*: Early environmental economics was based largely on the experience of a single country, the United States, where intense debate accompanied the formulation of policies for industrial air and water pollution control in the 1970s. Their involvement in the debate gave American economists a headstart in environmental economics. In essence, environmental economics was cast in an American mold [13]. There have been dramatic changes. Today, EPA employs economic analyses to improve the effectiveness of its environmental policies. The traditional approach in the United States is to regulate environmental pollution via pollution standards and impose pollution tax on the amount of pollution that companies emit into the environment.
- *China:* Much attention has been given to China's quick rise in the world. When more than one billion people join the global community, China's impact reaches far beyond its borders. The importance of the economy and the environment are reflected in China's rise to the world's 2nd largest economy and largest CO_2 emitter. China has sixteen of the world's twenty most polluted cities [14].
- *United Kingdom*: In January of 2018, the UK Government published its 25 Year Environment Plan. The Plan spells out the government's goals for environmental policy and the action it plans for improvement. To be concrete, it aims to deliver cleaner air and water, reduce risk of harm from floods and drought, use resources more sustainably and efficiently, enhance the beauty and engagement with the natural environment, and protect natural capital. The environment (or natural capital) underpins all of these things. Natural capital is an economics-based concept of our natural assets such as forests, rivers, land, minerals, and oceans [15].
- *Australia:* Australia is the smallest continent and one of the largest countries on the world. Environmental issues have become an important component in the political landscape in Australia. They play a significant role in election campaigns at both state and federal levels. The trade-offs involved in environmental policy have generated substantial opposition to it and served to underline the prominence of environmental questions in Australian politics. In spite of importance attached to "the environment," public debate is still characterized by "feel good" announcements by politicians, sensationalist media reporting, and deceptive claims by some environmental interest groups. It is becoming evident that the economic approach to the environment often forms the focal point of dealing with appropriate policies for the environment [16].
- *Austria:* Environmental economics is steeped in standard neoclassical theories of efficiency and welfare economics. These theories have been rejected by Austrian economists as conceptually unsound and as yielding analysis that does not reflect the real world. Thus, Austrian economics lacks a formalized theory of environmental economics. But by bringing together Austrian concepts of costs and the praxeological foundations of economics, one discovers a unique

perspective on pollution and the role of property rights in solving environmental problems. From this perspective, the process of production, exchange, and consumption in a strictly voluntary setting cannot be free of the kinds of inefficiencies generated by the negative externalities. This brings to bear a uniquely Austrian perspective on both the positive and normative analysis of environmental problems. The focus of the Austrian approach to environmental economics is conflict resolution [17].

8.5 BENEFITS

In essence, environmental economics examines how economic activity and policy affect the environment. It has contributed a great deal to our understanding the nature of environmental problems. The applications of economic theory can relate to any environmental issue and economic sector: agriculture, air quality, biodiversity, renewable energy, aquaculture production, human health, forestry, land-use, pollution, recreation, recycling, urban development, water supply, water quality, waste management, etc. Government policies increasingly require environmental economists to appraise and evaluate policy [18].

Environmental economics has enjoyed great success in recent decades. In other words, applying economic theory to environmental issues has been a booming endeavor. However, some see the boom as a mixed blessing. Environmental economics will help one to understand some important environmental issues such as climate change policy, recycling policy, traffic congestion charging, depletion of natural resources, and the tradeoff between cleaner environment and economic costs.

Market data can shed light on the benefits of environmental economics. Environmental economics offers powerful tools for the diagnosis of environmental problems and the design of policy solutions to them. There is a close connection between human health and environmental economics. Environmental economists have also helped measure the damages and the costs of pollution control. Embracing environmental economics will help us make better choices and long-term decisions. This can improve our approach to making sure we have clean air and water and that we are protected from risks such as floods [15].

8.6 CHALLENGES

The challenges facing environmental economics are many and pressing. The discipline has been severely constrained by the non-observability of some of its most fundamental concepts. Two main challenges to environmental economics are its transnational nature and its impact on various moving parts of a society. The global nature of environmental issues has led to the formation of organizations like the International Panel on Climate Change (IPCC), which organizes annual forums for Heads of State to negotiate international environmental policies. In the United States and some other nations, policy proposals stemming from environmental economics tend to lead to contentious political debate, making it difficult to develop environmental policies [19]. Examining environmental health issues through the eyes of microeconomics is a daunting task. It is very difficult to estimate the damages from environmental pollution with any reasonable degree of confidence. The fundamental economic challenge is one of measurement: how do we measure the value (as opposed to the price) of a good or service? What is the cost of oil spill on the ocean? What is the social cost of smoking? Can taxes be used to control pollution? How do you quantify the pollution that businesses emit into the environment? Uncertainty is ubiquitous in environmental economics. Since the real world is considerably uncertain, incorporating uncertainty into policy design is challenging [20]. There are challenging times ahead for environmental economics as new areas of research are arising.

8.7 CONCLUSION

It is not possible to describe human behavior independently from the environmental systems that sustain their existence. This problem can be overcome only through the inclusion of normative concepts like rights or justice. Environmental economics is a branch of economics that deals with the relationship between the economy and the environment. It studies economic effects and their consequences for human well-being and the natural environment. The field has contributed substantially to our understanding of the nature of environmental problems [21]. The fundamental principles of environmental economics are important in achieving an integrated sustainability science.

More information on environmental economics can be found in [8,14, 22-69]. More information is also available in the following journals devoted to it:

- *Journal of Environmental Economics and Management*
- *Journal of Environmental Economics and Policy*
- *Journal of Development Economics*
- *Journal of the Association of Environmental and Resource Economists*
- *Environmental and Resource Economics*
- *Review of Environmental Economics and Policy*

REFERENCES

[1] B. K. Jack, "Environmental economics in developing countries: An introduction to the special issue," *Journal of Environmental Economics and Management,*

vol. 86, November 2017, pp. 1-7.

[2] G. Roos, "Environmental economics in the chemical processes industry," *Water Science and Technology*, vol. 39, no. 10-11, 1999, pp. 25-30.

[3] "Economics and the environment,"

http://tracoecobalt.asu.lt/files/outgrowth/books/lat_en/Chapter_1/chapter_1.1.htm

[4] S. Beder, "Environmental economics and ecological economics: the contribution of interdisciplinarity to understanding, influence and effectiveness," *Environmental Conservation,* vol. 38, no.2, June 2011, pp. 140-150.

[5] M. N. O. Sadiku, O. D. Olaleye, and S. M. Musa, "Environmental economics: A primer," *International Journals of Advanced Research in Computer Science and Software Engineering,* vol. 9, no. 7, July 2019, pp. 36-39.

[6] "Environmental economics," *Wikipedia*, the free encyclopedia,

https://en.wikipedia.org/wiki/Environmental_economics

[7] "Environmental vs. ecological economics,"

https://www.economicseducation.org/starter-pack

[8] K. V. Pavithran, *A Textbook of Environmental Economics.* New Age International, 2008.

[9] "Sustainable development,"

https://www.researchgate.net/figure/Sustainable-development_fig1_329218259

[10] Department of Environmental Affairs and Tourism, "Environmental economics,"

www.deat.gov.za

[11] "What is an environmental economics"? Unknown Source.

[12] "What is an environmental economist?"

https://www.environmentalscience.org/career/environmental-economist

[13] O. Braadbaart, "American bias in environmental economics: Industrial

pollution abatement and 'incentives versus regulations'," *Environmental Politics*, vol. 7, no. 2, 1998, pp. 134-152

[14] S. Managi and S. Kaneko, *Chinese Economic Development and the Environment:*

New Horizons in Environmental Economics Series. Cheltenham, UK: Edward Elgar. 2009.

[15] J. Curnow, "Environmental economics in UK environmental policy: Defra's 25 Year Environment Plan," *Journal of Environmental Economics and Policy,* vol. 8, no.4, 2019, pp. 353-358.

[16] B. Dollery and J, Wallis, "A cautionary note on environmental economics," *Australian Journal of Social Issues,* vol. 32, no. 3, August 1997, pp. 299-305.

[17] R. Cordato, "An Austrian theory of environmental economics," September 2005,

https://mises.org/library/austrian-theory-environmental-economics

[18] K. Willis, E. Ozdemiroglu, and D. Campbell, "Environmental economics and policy," *Journal of Environmental Economics and Policy*, vol. 1, no.1, 2012, pp. 1-4,

[19] J. Chen, "Environmental economics," April 2019,

https://www.investopedia.com/terms/e/environmental-economics.asp

[20] R. S. Pindyck, "Uncertainty in environmental economics," *Review of Environmental Economics and Policy*, vol. 1, no. 1, Winter 2007, pp. 45–65.

[21] S. Vanderheiden, "Missing the forest for the trees: Justice and environmental economics," *Critical Review of International Social and Political Philosophy,* vol. 8, no. 1, 2005, pp. 51-69.

[22] B. C. Lin and S. Zheng (eds.), *Environmental Economics and Sustainability.* John Wiley & Sons, 2017.

[23] P. M. Schwarz, *New Energy Economics Book.* Routledge, 2018.

[24] H. Folmer and E. van Ierland (eds), *Valuation Methods and Policy Making in Environmental Economics.* Amsterdam: Elsevier Science Publishers, 1989.

[25] A. Markandya and J. Richardson, *Environmental Economics: A Reader.* New York, NY: St. Martins Press, 1993.

[26] K. G. Mäler and J. R. Vincent (eds.), *Handbook of Environmental Economics: Valuing Environmental Changes.* Amsterdam: Elsevier/North-Holland, volume 2, 2005.

[27] V. S. Ganesamurthy, *Environmental Economics in India.* New Delhi, India: New Century Pub., 2009.

[28] J. F. Shogren. (ed.), *Experiments in Environmental Economics.* Ashgate Publishing Co., 2003.

[29] K. Aravossis et al. (eds.), *Environmental Economics Environmental Economics and Policy.* WIT Press, 2006.

[30] C. Tisdell, *Environmental Economics: Policies for Environmental Management and Sustainable Development.* Edward Elgar Publishing Limited, 1993.

[31] A. Markandya and J. Richardson (eds.), *The Earthscan Reader in Environmental Economics. London: Earthscan Publications Limited, 1992.*

[32] D. Helm (ed.), *Economic Policy Towards the Environment.* Oxford: Blackwell, 1991.

[33] D. J. Phaneuf and T. Requate, *A Course in Environmental Economics: Theory, Policy, and Practice.* Cambridge, UK: Cambridge University Press, 2017.

[34] B. Field and M. K. Field, *Environmental Economics: An Introduction.* McGraw-Hill, 7th Edition. 2017.

[35] M. Boman, R. Brännlund, and B. Kriström, *Topics in Environmental Economics.* Springer, 1999.

[36] P. Dasgupta, S K. Pattanayak, and V. Kerry Smith (eds.), *Handbook of Environmental Economics.* North Holland, Volume 4, 2018.

[37] H. Wiesmeth, *Environmental Economics: Theory and Policy in Equilibrium.* Springer 2012.

[38] A. Markandya et al., *Environmental Economics Forsustainable Growth: A Handbook for Practitioners.* Cheltenham, UK: Edward Elgar, 2002.

[39] S. Managi (ed.), *The Routledge Handbook of Environmental Economics in Asia.* Routledge, 2019.

[40] T. H. Tietenberg and L. Lewis, *Environmental Economics and Policy.* Addison-Wesley Professional, 2009.

[41] A. Gilpin, *Environmental Economics: A Critical Overview.* John Wiley & Sons, 2000.

[42] S. Managi and K. Kuriyama, *Environmental Economics.* Routledge, 2016.

[43] M. Karpagam, *Environmental Economics A Textbook.* Sterling Publishers, 2019.

[44] R. Perelet et al., *Dictionary of Environmental Economics.* London, UK: Routledge, 2001.

[45] I. Hodge, *Environmental Economics: Individual Incentives and Public Choices.* Springer, 1995.

[46] J. Asafu-Adjaye, *Environmental Economics for Non-Economists: Techniques and Policies for Sustainable Development.* World Scientific, 2nd Edition, 2005.

[47] T. H. Tietenberg and L. Lewis, *Environmental Economics: The Essentials.* Taylor & Francis, 2019.

[48] T. Tietenberg and L. Lewis, *Environmental Economics & Policy: Global Edition.* Pearson, 6th edition, 2013.

[49] G. Squires, *Urban and Environmental Economics: An Introduction.* Routledge, 2012.

[50] R. N. Stavins et al., (eds.), *Economics of the Environment: Selected Readings.*

Edward Elgar, 7th edition, 2019

[51] D. J. Thampapillai, *Environmental Economics: Concepts, Methods, and*

Policies. Oxford University Press, 2002.

[52] William K. Jaeger, *Environmental Economics for Tree Huggers and Other Skeptics.*

Island Press, 2005.

[53] R. K. Turner, D. W. Pearce, and I. Bateman, *Environmental Economics: An Elementary*

Introduction. Johns Hopkins University Press, 1993.

[54] R. Benelmir (ed.), *Energy-Environment-Economics.* Nova Science Publishers, 2014.

[55] W. Buchholz and D. Rübbelke, *Foundations of Environmental Economics.*

Springer International Publishing, 2019.

[56] D. A. Anderson, *Environmental Economics and Natural Resource Management.* London,

UK: Routledge, 4th edition, 2013.

[57] A. M. Hussen, *Principles of Environmental Economics.* Routledge, 2nd edition, 2004.

[58] I. J. Bateman, A. A. Lovett, and J. S. Brainard, *Applied Environmental Economics: A GIS Approach to Cost-Benefit Analysis.* Academic, 2005.

[59] D. Chapman, *Environmental Economics - Theory, Application, and Policy.*

Reading, MA: Addison Wesley, 2000.

[60] J. Thampapillai and J. A. Sinden. *Environmental Economics – Concepts, Methods and Policies.* OUP Australia & New Zealand, 2nd edition, 2013.

[61] K. G. Mäler and J. R. Vincent (eds.), *Handbook of Environmental Economics: Valuing Environmental Changes.* Amsterdam: Elsevier/North-Holland, volume 2, 2005.

[62] N. Hanley, J. F. Shogren, and B. White, *Environmental Economics in Theory and Practice.* Palgrave Macmillan, 2nd edition, 2007.

[63] C. Kolstad, *Intermediate Environmental Economics: International Edition.* India: Oxford University Press, 2nd edition, 2011.

[64] R. K. Turner, D. Pearce, and I. Bateman, *Environmental Economics: An Elementary Introduction.* Hemel Hempstead, UK: Harvester Wheatsheaf, 1994.

[65] A. C. Fisher, *Resource and Environmental Economics.* Cambridge, UK: Cambridge University Press, 1981.

[66] J. Bowers, *Sustainability and Environmental Economics: An Alternative Text.* Harlow, UK: Addison Wesley Longman, 1997.

[67] S. Callan and J. Thomas, *Environmental Economics and Management: Theory, Policy and Applications.* Fort Worth: Dryden Press, 2nd edition, 2000.

[68] T. Tietenberg and L. Lewis, *Environmental & Natural Resource Economics.* Pearson Education, 9th edition, 2009.

[69] R. Perman et al., *Natural Resource and Environmental Economics.* Pearson Education, 3rd edition, 2003.

CHAPTER 9

ENVIRONMENTAL MANAGEMENT

"Management, at every level, is about the effort to frame challenges, define end states, and allocate resources to navigate between them." - Steven Sinofsky

9.1 INTRODUCTION

Humans have been the primary agents altering the conditions of our planet. The rapid development of the industrial sector and globalization have resulted in massive environmental degradation in the form of energy consumption, toxic waste, and pollution.

The amount of E-waste has grown steadily. (E-waste refers to electronic products nearing the end of their "useful life"). Prevention of E-waste generation, recycling capacity, reuse process, treatment techniques, as well as the legislation of E-waste are necessary. High exposure of hazardous materials is dangerous to human life. Therefore, a sustainable environmental management (EM) of E-waste by governing regulations is important.

The world's renewable resources such as water, soil, forests, and fish are under increasing pressure. These pressures have led to the idea of "sustainable development," which is basically the development that meets the needs of the present without compromising the ability of future generations to meet their own needs [1]. Corporations worldwide are embracing environmental protection and pursuing proactive environmental management due to pressures from governments, customers, employees, and competitors. They now recognize that environmentally responsible management is a requirement for doing business.

Many organizations now show their care for the environment by reducing their environmental footprint. To assure a cleaner environment, government regulations are becoming more stringent and customers are becoming more demanding. Cleaner production takes strategic, comprehensive, and preventive measure. Environmental management has become an important component of corporate social responsibility and a company's strategic agenda [2].

This chapter is intended to provide an introduction to EM so that beginners can understand environmental management, its increasing importance, and new developments. It begins by discussing what environmental management is all about. It explains the characteristics of environmental management. It covers environmental management system by which it is implemented. It also covers environmental performance evaluation and provides some applications of environmental management. It addresses environmental management accounting. It gives global perspectives on environment management. It highlights the benefits and challenges of environment management. The final section concludes with comments.

9.2 CONCEPT OF ENVIRONMENTAL MANAGEMENT

Environmental management (EM) is a multidisciplinary area that is concerned with the management of human activities and their impacts on the natural environment. It is basically about making decisions on the use of natural resources. It involves pressing issues of justice and survival. Environmental managers consist of a diverse group of people including academics, policy-makers, non-governmental organization (NGO) workers, company employees, civil servants and other individuals or groups who desire to control the direction and pace of development.

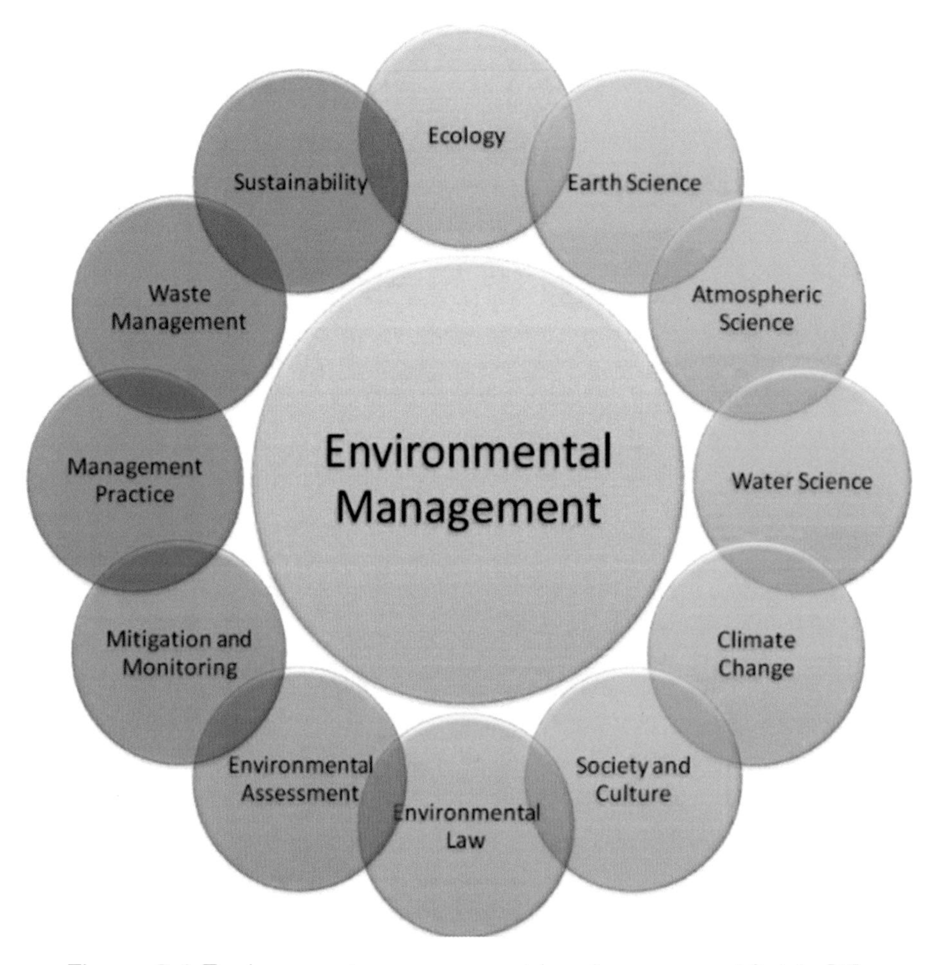

Figure 9.1 Environment management involves several fields [3].

Environmental management is a new branch of management that emerged in the 1980s and adopted by a number of firms in the 1990s. As shown in Figure 9.1, EM involves several areas such as environment, ecology, waste management, sustainability, and law [3]. Environmental management is mainly implemented by industrial sectors, such as manufacturing, paper, steel, oil, and chemicals, which are constrained by regulations to reduce pollution. Several enterprises, such as GM, BMW, IBM, and Hilton, have used improved environmental performance as a means of increasing market share and differentiating their products from those of their competitors. Such environmental leaders have been reported to have at least five similar internal characteristics [4]:

- Improvement of eco-efficiency (e.g., reduction of product wastes) over time.
- Performance of proactive measures to control environmental risks.
- Creation of business value through environmental strategies that enhance their competitiveness.
- General improvement of relations with key stakeholders (e.g., socially responsible marketing).
- Formation of business processes that increase innovation and identify competitive advantages.

9.3 CHARACTERISTICS OF ENVIRONMENTAL MANAGEMENT

A coordinated approach is necessary to solve environmental problems. The three pillars of EM are sustainability, corporate social responsibility, and corporate governance.

- *Sustainability:* This is often regarded as development that meets present needs without compromising the ability of future generations to meet their own needs. A sustainable development is one that guarantees a durable satisfaction of the human needs and increases the quality of life. As organizations become more aware of the importance of actions that can lead to greater sustainability, the process of organizational learning and change takes place. Environmental management practices are related to environmental reputation when green advertising is used to publicize information about the environmental efforts of manufacturers [5].
- *Corporate social responsibility* (CSR): Social responsibility is the principle that requires organizations to contribute to the welfare of society (such as focusing on poverty relief, education, or the betterment of the community) and not solely be interested in maximizing profits. CSR carries the idea that companies and businesses can no longer act as isolated economic entities operating in detachment from broader society. CSR is a company's commitment to manage the social, economic, and environmental effects of its operations responsibly. It is an important issue in business practice across all industries. Practicing corporate social responsibility makes a company to be conscious of the kind of impact it is making on all aspects of society including economic, social, and environmental. Companies that engage in CSR focus on the triple bottom line: people, planet, and profit [6].
- *Corporate Governance*: Every organization must make environmental management an organizational priority in order for EM to succeed. EM must be incorporated in all aspects of the organization. Environmental responsibility is an integral part of every employee's function. Everyone must be included in planning, awareness training, and other motivational programs. The company's environmental policy must be defined by top management. It must include the environmental impacts of the company, its commitment to environmental compliance, reduction or control of pollution, and must be publicly available [7]

The incorporation of EM practices in company operations can also be classified in three ways: (a) the planning, organization, direction, and control of behavioral and corporate aspects of EM; (b) environmental improvements of products and production processes; and (c) practices for internal and external communication of the company's environmental improvement initiatives and results [8].

9.3 ENVIRONMENTAL MANAGEMENT SYSTEM

The management of any facility requires a watchful eye to ensure regulations and procedures are followed and safety procedures are properly performed. All organizations somewhat consider environmental issues in their overall management processes in support of their missions in an environmentally, economically, and fiscally sound manner. An environmental management system (EMS) refers to a globally accepted organizational management practice that allows an organization to strategically address its environmental issues. It is a way of incorporating environmental thinking into an organization's daily operations [9]. The implementation of environmental management is done through the use of an environmental management system.

Today, companies and organizations cannot ignore environmental issues. Governments, stakeholders, and customers are increasingly expecting organizations to demonstrate their commitment to managing their impacts on the environment. Environmental management is a proactive systematic approach to finding practical ways for saving natural resources (e.g. land, water, and materials), energy, and reducing negative environmental impacts. One means of bringing about improved environmental performance is through the adoption of environmental management system (EMS). EMS is implemented to guide how a company manages its potential environmental impact and ensure environmental compliance with legislation.

An environmental management system (EMS) is a management system that focuses on incorporating environmental considerations in business practices. It is a framework that helps an organization to achieve its environmental goals through consistent review, evaluation, and focusing on the reduction of its environmental footprint. An EMS encourages an organization to continuously improve its environmental performance [10].

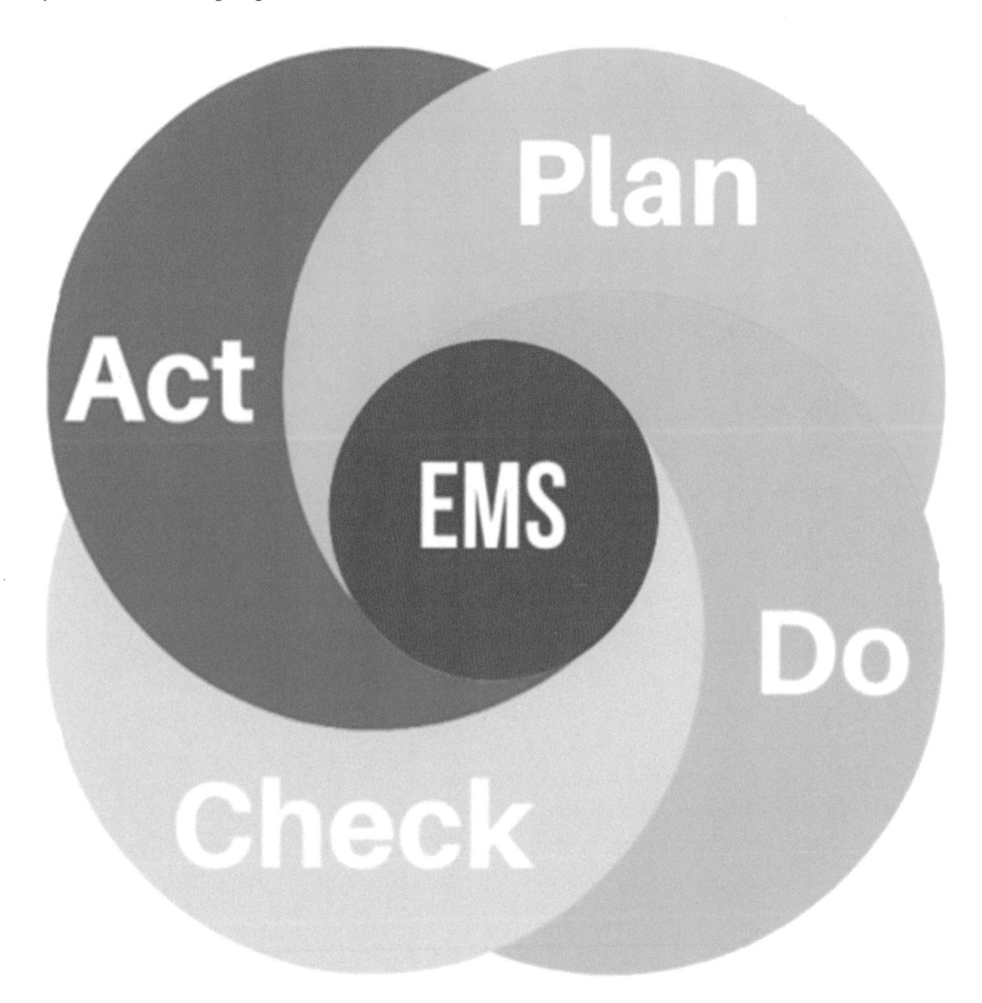

Figure 9.2 The "Plan-Do-Check-Act" cycle for environmental management system [11].

EMS is a system, a documentation of procedures, a plan for continual improvement, a tool for setting goals for the future, and a way to involve your entire operation in better environmental operations. A simple model that may be useful starting point for creating your business' environmental management system is the "Plan-Do-Check-Act" cycle, illustrated in Figure 9.2 [11]. The most commonly used standard for an EMS is the one developed by the International Organization for Standardization (ISO) for the ISO 14001 standard. More than 300,000 organizations worldwide have certified to ISO 14001. The five major stages of an EMS, as defined by the ISO 14001 standard, are shown in Figure 9.3 [12].

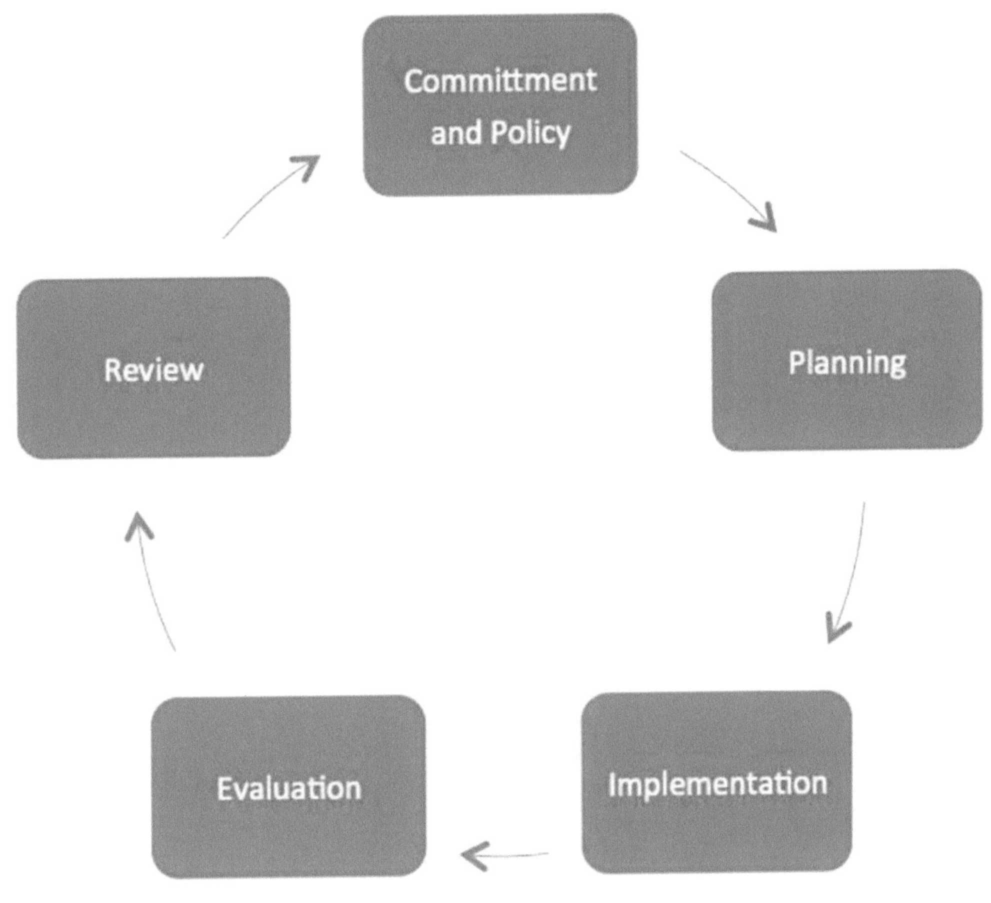

Figure 9.3 Five stages of EMS [12].

An environmental management system (EMS) does the following [13]:

- Serves as a tool, or process, to implement EM and improve environmental performance.
- Provides a systematic way of managing an organization's environmental affairs.
- Addresses immediate and long-term impacts of its products, services and processes on the environment.
- EMS assists with planning, controlling, and monitoring policies.
- Creates environmental buy-in from management and employees and assigns accountability and responsibility.
- Sets framework for training to achieve objectives and desired performance.
- Helps understand legislative requirements to better determine a product or service's impact, significance, priorities, and objectives.
- Focuses on continual improvement of the system and a way to implement policies and objectives to meet a desired result.
- Encourages contractors and suppliers to establish their own EMS.
- Facilitates e-reporting to federal, state, and local government environmental agencies.

9.4 ENVIRONMENTAL PERFORMANCE EVALUATION

Environmental management is a proactive systematic approach to finding practical ways for saving natural resources (e.g. land, water, and materials), energy, and reducing negative environmental impacts. Improving the environmental performance of enterprises is important, irrespective of their size. One means of bringing about improved environmental performance is through the adoption of EMSs.

As mentioned earlier, EMS is a structured framework under which an entity can manage environmental impacts. This is a useful tool for organizations that intend to integrate environment management in their corporate policy. It is a tool to measure environmental performance to validate continuous improvement. EMSs such as IS014001 and EMAS may be major growth areas for the accredited certification. The voluntary adoption of EMS by companies has become a vital supplement to mandatory environmental policies [14]. EMS has been adopted by organizations to improve product quality, reduce production cost, and improve reputation or corporate image.

Environmental management will be mainly guided by International Seabed Authority (ISA) rules, regulations, procedures and guidelines. ISO 14000 standard brings environmental management into its management so that all employees are aware of environmental issues and clearly understand their responsibilities.

The technique of life-cycle assessment (LCA) is one of the tools that organizations have at their disposal to help assess the environmental impact of their products or services. LCA is a technique which is becoming widely adopted since it can provide a more scientific and objective platform on which to base an evaluation. LCA focus is evaluation of the environmental impact of a product, while EMS is organization's management system. The relationship between EMS and LCA is shown in Figure 9.4 [15].

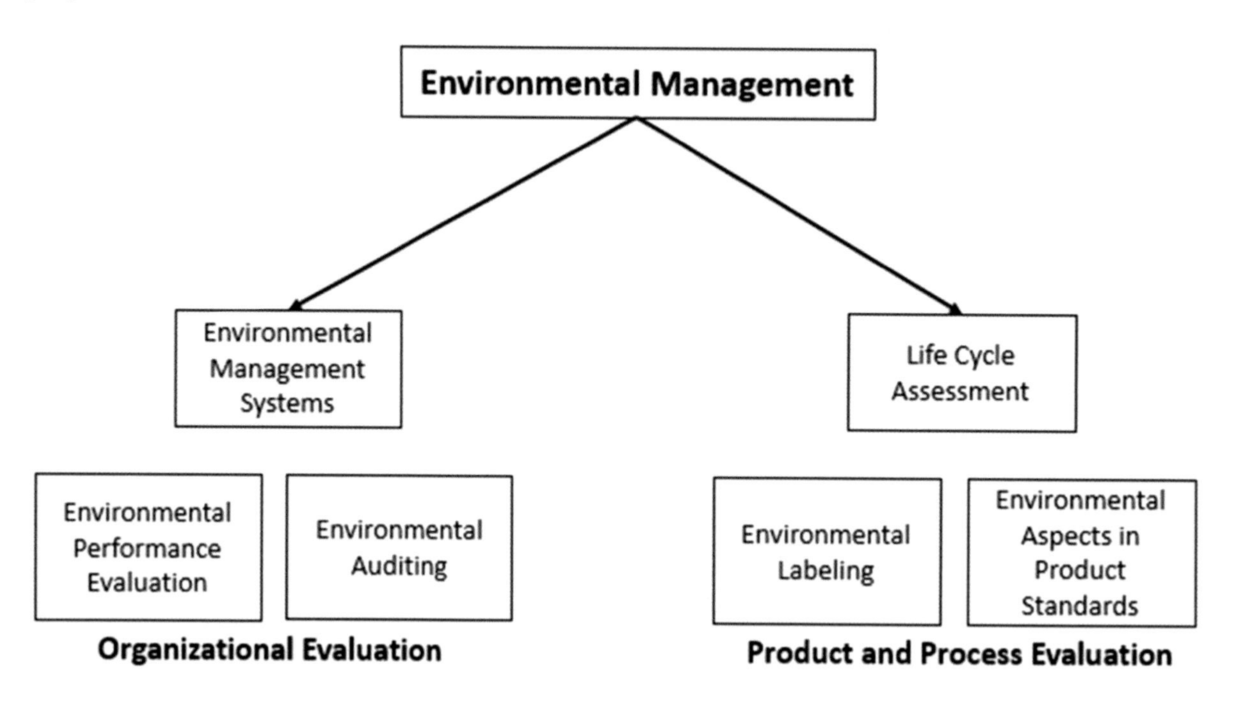

Figure 9.4 The relationship between EMS and LCA [15].

9.5 APPLICATIONS OF ENVIRONMENTAL MANAGEMENT

This section provides some applications of environmental management. Environmental issues have become central to the competitive success of many organizations.

- *Environmental Accounting:* Accounting plays a crucial role in ensuring environmental sensitivity. In organizations, accounting performs a fundamental role through both the internal decision making and the external reporting on a firm's performance. The social responsibility of a firm has been highlighted in recent debate on the accountability function of accounting [16]. More will be said about environmental management accounting in the next section.

- *EM in the Hotel Industry:* Hotels are a central component of the hospitality industry, contributing significantly to local economies. Environmental management in hotels reflects a paradigm shift in the industry from mass tourism to sustainable tourism. The awareness of the quests about hotels' sustainability practices has been growing. Chain hotel companies have developed their own environmental reporting systems for franchised hotels in an effort to gauge and improve their sustainability. Many green practices in the hotel industry (e.g., recycling program) are relatively simple measures. Promoting green hotel measures can be mainly divided into three stages: process efficiency, people efficiency, and system efficiency [17,18].
- *EM in Tourism Industry:* Tourism is of great economic and social benefit to many countries. Environmental issues are becoming important to the tourism industry and how these issues are handled will have far-reaching effects on how tourism operates. Accommodation is the largest sub-sector of the tourism industry undeniably and has the greatest impacts on the environment [19].
- *EM in Food industry:* More researches are devoted to environmental impact of food production industry. Farmers, government policy makers, donor organization officials, researchers, scientists, and other participants in aquaculture realize that environmentally responsive production technologies need to be applied [20].
- *Enterprise Environmental Management:* Enhancing corporate or enterprise environmental management of human resources has become a key factor of good corporate environmental performance and sustainable development. This should include strengthening the enterprise environment management, professional human resources construction, and the environmental consciousness of all the staff.

Other areas of applications include environmental monitoring, marine mining, local government, industrial goods supply chains, agriculture, port management, iron and steel industry, military training areas, construction industry, small and medium sized enterprises, and offshore mining.

9.6 ENVIRONMENTAL MANAGEMENT ACCOUNTING

Environmental management accounting (EMA) is environmental accounting which focuses mainly on providing management information for internal environmentally beneficial decision-making. It represents an approach which provides increase in material efficiency, decrease in environmental impact, and reduction in costs of environmental protection. It is a part of the management accounting that focuses on things such as the cost of energy, water, and waste disposal. Its major goal is to improve both economic and environmental performance of a company by utilizing both financial and non-financial information. EMA has been acknowledged to deliver several benefits to organizations [21].

Conventional national accounting ignores the exhaustion of environmental resources and the degradation of the environment. Critics have argued that gross domestic product (GDP) (as a measure of national income) overlooks the environmental factor and, therefore, needs to be revised to incorporate green accounting. Green or environmental accounting seeks to correct the aggregated indicator of national accounting, the GDP. It is a new branch of accounting that provides for accounting the environmental impact. EMA may be regarded as the next step in the evolution of management accounting. Figure 9.5 shows the four approaches to environmental accounting [22].

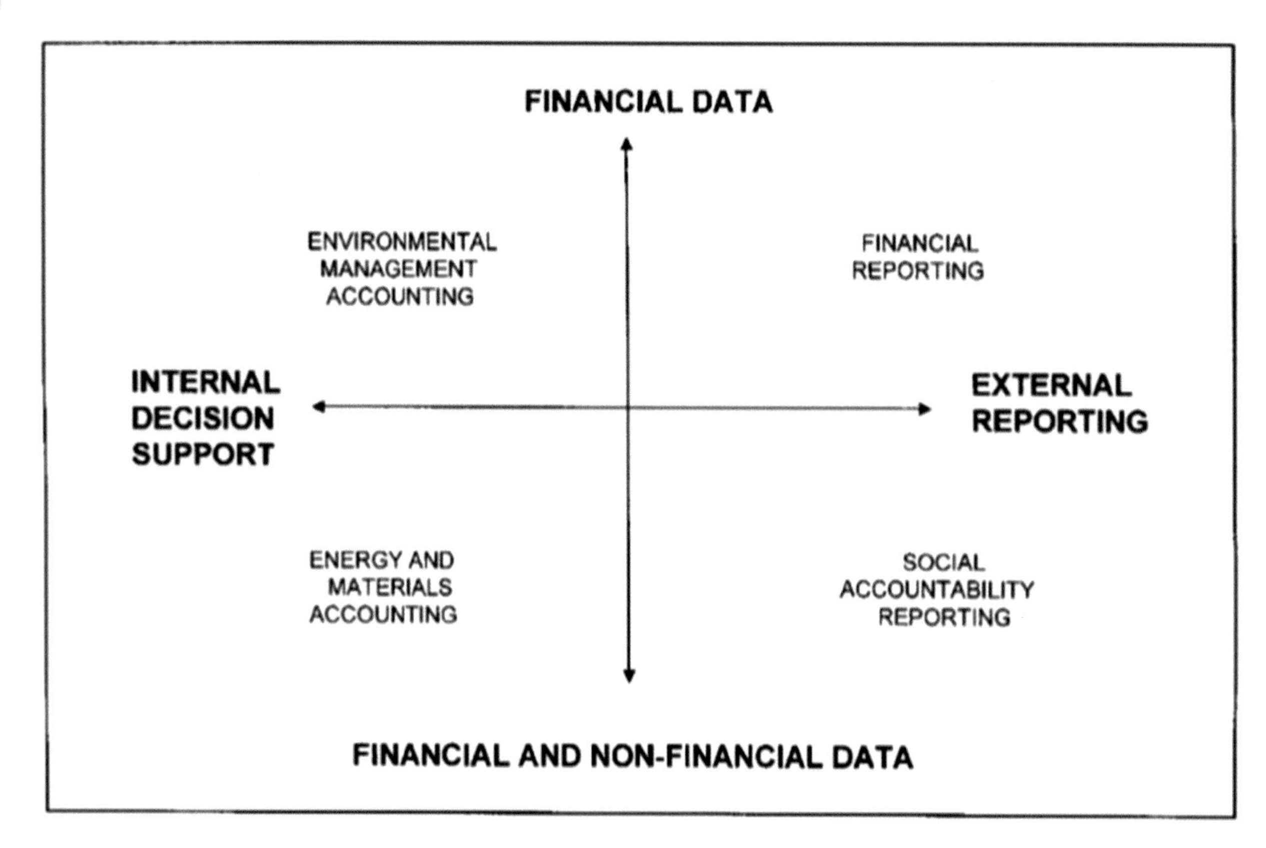

Figure 9.5 Four approaches to environmental accounting [22].

Environmental management accounting (EMA) has emerged in recent decades in response to the growth of environmental problems. EMA is a set of principles and approaches that use environmental information for internal decision-making by implementing appropriate environment-related accounting systems and practices. For example, waste reduction programs cannot be reliably estimated without using EMA tool. EMA is a subset of environmental accounting.

EMA deals with the internal decision making related to the environmental performance of an organization. It may be regarded as an application of conventional accounting that is concerned with the environmentally-induced impacts of companies. EMA is an important tool for integration of environmental considerations into decision-making of an organization. Although EMA may still be unknown concept to many, its principles are practiced by many organizations worldwide.

9.7 GLOBAL PERSPECTIVES ON ENVIRONMENT MANAGEMENT

Corporations in North America, Europe, Japan, and other nations are embracing environmental management due to pressures from governments, customers, employees, and competitors to assure a cleaner environment. Several environmental agencies worldwide are forging ahead with the concept. We now present the perspective of different nations on environmental management.

- *United States:* In the US, professional associations, scientific organizations, and other interest groups have increasingly championed environmental management. Companies in the US have attempted to get some regulatory relief by implementing an EMS that meets the ISO 14000 standard. The US Technical Advisory Group (US TAG) develops the US position on various ISO 14000 standards. The objective of the standards is to help companies develop a set of environmental management elements that will help to achieve their environmental goals. It also helps to facilitate the development of environmental management systems in various organizations [23]. For example, construction in the US makes a smaller contribution to hazardous waste generation than its share of GDP might suggest. This probably demonstrates that the US Environmental Protection Agency moves to regulate these environmental emissions more closely.

- *Hong Kong:* The promotion of environmental management has resulted in pressure demanding the adoption of proper methods to improve environmental performance across all industries including construction. Construction is not by nature an environmentally friendly activity. It is a major contributor to environmental pollution, which is typically classified as air pollution, waste pollution, noise pollution, and water pollution. The notion that environmental management results in many more costs than benefits is prevalent among contractors in Hong Kong. The contractors do not generally seem to consider that the implementation of environmental management will bring cost savings. The local industry has been promoting measures such as establishing waste management plans, reduction and recycling of wastes, providing in-house training on environmental management, and legal measures on environmental protection [24].
- *Zimbabwe:* The management of environmental resources in any nation should to be well coordinated. Effective coordination of environmental management institutions remains key to the achievement of sustainable water resources management in the south African country of Zimbabwe. For a long time, Zimbabwe has acknowledged the importance of environmental management and sustainability. Environmental management practitioners in the nation concur that the major environmental challenges include siltation, water pollution, deforestation, and water shortage. The Environmental Management Act (EMA) is the principal environmental management legislation in the country. It is the lead agency for environment management in Zimbabwe [25].
- *Iraq*: In the Iraq, petroleum products including fuel and lubricants are considered hazardous materials, known collectively as POL - petroleum, oil and lubricants. POL materials are used in large quantities. Some hazardous materials have been seized by coalition forces in Iraq due to their potential for use in homemade explosives. Storage and disposal of hazardous wastes across a large-scale theater of operations presents some logistical challenges. Hazardous waste treatment centers have been established at different locations to address hazardous waste streams [26].
- *China*: Since 1978, urbanization in China has been accelerated as a result of rapid economic development. Today, China faces crucial issues such as poor air quality, lack of adequate drinking water supplies, inappropriate waste disposal practices, transportation congestion, land subsidence, etc. An important aspect of Chinese urban environmental management is public participation. Public participation is carried out in several ways: prompt stopping of the behavior of pollution, reporting to the mass media, administrative complaint and justice litigation, conveying opinions and suggestions on environment through the People's Congress and the Political Consultative Conference. China needs to cultivate a comprehensive system of cultural values based on environmental ethics and morals. It also needs assistance and in environmental management from environmentally successful nations and organizations [27].
- *Canada:* Rural areas in Canada are both landscapes of production and hosts to conservation and biodiversity-protection initiatives. Its biosphere reserves are regarded protected areas. Canadian manufacturing facilities have undertaken environmental initiatives as a result of pressures arising from the buyers of their products. An increasing number of business initiatives attempt to change corporate culture and management practices by introducing environmental

management. This should encourage companies that have no or very few environmental practices to adopt more environmental practices [28].

- *Kenya:* The world's worst environmental concerns are in developing nations. Governments have been compelled to establish laws that demand that businesses should cut down on pollution. Droughts and intense flooding continue to occur in Kenya every year since 2000. Predictions suggest that an increasing amount of land will be made unarable as a result of more erratic rainfall, while the impacts of climate change on livelihoods have been particularly severe in the country. Desertification is growing in the arid and semi-arid areas. For about 20 years, Kenya has witnessed the development of new state organizations with a broader environmental mandate. Assistance to environmental and natural resources management increased steadily [29].
- *United Kingdom:* The United Kingdom's environment has improved substantially over the past five decades. The imperatives of compliance with legislation, cost and risk reduction, and sustainability continue to drive the action plans of UK. In the UK, environmental regulators have traditionally sought the voluntary compliance of businesses. Its environmental regulators will soon have administrative penalties (e.g. civil fines and license suspensions) at their disposal for protecting the environment. It seems that British businesses and other organizations are less enthusiastic about the European Community's Eco-Management and Auditing Scheme (EMAS) [30].

9.8 BENEFITS

Environmental management has the potential to play a crucial role in the financial performance of a company. Its positive impacts include economic development, generation of new jobs, and an increase in specialized labor. Companies are beginning to recognize the possible competitive advantages associated with environmental awareness.

Environmental management is about predicting future environmental changes, maximizing human benefit, and minimizing environmental degradation. Sound environmental management safeguards nature from potentially damaging business operations, constructing a win-win situation that benefits both the economy and the environment.

Other key benefits of incorporating an environmental management system in your business are summarized as follows [31]:

- Enforces sustainable action and provides sustainable profits, for long-term business success
- Establishes a marketing advantage
- Reduces the frequency of environmental incidents
- Improves your businesses reputation
- Is a win-win approach that reduces operating costs and improves profitability
- Attracts shareholders and investors
- Improves your regulative performance
- Protects the environment from the effects of manufacturing by-products
- Protects your business from noncompliance fines and penalties
- Ensures a holistic approach to environmental impacts
- Focuses on only critical aspects and processes
- Makes use of time-tested, mature approaches recognized worldwide
- Establishes positive relationships with regulators

9.9 CHALLENGES

A major challenge presented by the environment to society and business is due to the managerial complexity of the issues it raises. Complexity arises because the environment is interpreted and sensed through a set of human constructs - politics, economics, science, culture, and religion. The complexity has impacts for technology, management techniques, organizational structures as well as individual and organizational values [32].

It makes a concise definition of EM elusive. The difficulty of managing environment issues tempts many enterprises to ignore or undermanage those issues, which is not prudent. There are significant challenges to the implementation of environmental practices among small and medium-sized enterprises (SMEs), which make up the vast majority of businesses. This is due to low awareness, economic barriers, and limited business support. Knowledge and best practice in EM are constantly changing.

Only large corporations can afford to have their own environmental department with diversified competencies. It takes time for new professional disciplines to be accepted by industry, to be recognized by the society, and to receive accreditation. Decision makers at all levels have become dissatisfied with narrowly-focused, incremental, and disjointed environmental management.

Other barriers that impede EMSs implementation or the adoption of EMSs include [33]:

- Lack of management and/or staff
- Inadequate technical knowledge and skills
- Perception of high cost for implementation and maintenance
- Lack of government legal enforcement
- Increase in management and operation costs
- Lack of trained staff and expertise
- Lack of client support
- Lack of sub-contractor co-operation
- Lack of supplier co-operation
- Difficult co-ordination of environmental performance among multi-tier subcontractors
- Lack of working staff support
- Time-consuming for improving environmental performance
- Change of existing practice of company structure and policy
- Increase in documentation workload
- Lack of tailor-made training on environmental management
- Lack of technological support within organization

9.10 CONCLUSION

Environmental issues have become an important element of environmental management and strategies, compelling many companies to be more environmentally responsible. The environment and the economy are interdependent and inseparable. Environmental management may be regarded as the management of human impacts on the environment. As an explicit function in industry, environmental management has expanded considerably during the last two decades due to increasing corporate concern for the environment and its management.

Implementing an EMS, such as ISO 14001, is no longer an option for many companies. Environmental management has become a major concern for organizations, governments, and citizens. Enterprise and corporate environmental management attention is gradually moving from clean technologies and pollution prevention to green product development.

Every professional should endeavor to have a basic understanding of the technical and scientific terms related to the environment to be able to make smart, sound decisions. It should be noted that environmental improvement is a never-ending process. More information about EM can be found in the books in [34-60] and the following journals devoted to it:

- *Environmental Management*
- *Journal of Environmental Management*
- *Journal of Environment & Development*
- *Journal of Social and Environmental Management*
- *Journal of Environmental Economics and Management*
- *Journal of Environmental Planning and Management*
- *Management of Environmental Quality: An International Journal*
- *Journal of Environmental Assessment Policy and Management*
- *Journal of Indian Association for Environmental Management*
- *Indonesian Journal of Environmental Management and Sustainability*

REFERENCES

[1] R. Sullivan, "Being sustainable...be specific," *Engineering Management Journal,* October 2002, pp.220-225.

[2] M. N. O. Sadiku, O. D. Olaleye, and S. M. Musa, "Environmental management: A primer," *International Journal of Trend in Research and Development,* vol. 6, no. 3, May- Jun. 2019, pp. 105-107.

[3] "Online certificate in environmental management programme at UCD," May 2016.

http://www.engineersjournal.ie/2016/05/02/online-certificate-environmental-management-programme-ucd/

[4] L. S. Traves, "Double due diligence: Incorporating strategic environmental management in environmental assessments for industrial mergers and acquisitions,"

Environmental Practice, vol. 4, no. 4, December 2002, pp. 197-209.

[5] M. F. A. Goosen, "Environmental management and sustainable development,"

Procedia Engineering, vol. 33, 2012, pp. 6-13.

[6] M. N. O. Sadiku, S. R. Nelatury, and S.M. Musa, "Corporate social responsibility: A primer," *Journal of Scientific and Engineering Research,* vol. 6, no. 3, 2019, pp. 35-40.

[7] J. S. Waters, "Environmental management systems,"

http://www.kdheks.gov/environment/download/EMS.pdf

[8] J. A. P. de Oliveira and C. J. C. Jabbou, "Environmental management, climate change, CSR, and governance in clusters of small firms in developing countries: Toward an integrated analytical framework," *Business & Society,* vol. 56, no. 1, 2017, pp. 130 –151.

[9] M. N. O. Sadiku, O. D. Olaleye, and S. M. Musa, "Environmental management system: A tutorial," *International Journal of Trend in Research and Development,* vol. 6, no. 3, May - June 2019, pp. 136-139.

[10] "Learn about environmental management systems,"

https://www.epa.gov/ems/learn-about-environmentalmanagement-

systems

[11] https://www.supplychainschool.co.uk/topics/sustainability/environmental-management/

[12] "Why do you need an environmental management system?"

https://www.glion.edu/blog/needenvironmental-management-system/

[13] "Environmental management system," *Wikipedia,* the

free encyclopedia

https://en.wikipedia.org/wiki/Environmental_management_system

[14] M. Frondel, K. Krätschell, and L. Zwick, "Environmental management systems: Does certification pay?" *Economic Analysis and Policy,* vol. 59, 2018, pp. 14-24.

[15] S. A. Melnyka, R. P. Sroufeb, and R. Calanton, "Assessing the impact of environmental management systems on corporate and environmental performance,"

Journal of Operations Management, vol. 21, 2003, pp. 329–351.

[16] E. Bracci and L. Maran, "Environmental management and regulation: Pitfalls of environmental accounting?" *Management of Environmental Quality: An International Journal*, vol. 24, no. 4, 2013, pp. 538-554.

[17] J. Park, H. J. Kim, and K. W. McCleary, "The impact of top management's environmental attitudes on hotel companies' environmental management," *Journal of Hospitality & Tourism Research*, vol. 38, no. 1, February 2014, pp. 95-115.

[18] C. H. Yen, C. Y. Chen and H. Y. Teng, "Perceptions of environmental management and employee job attitudes in hotel firms," *Journal of Human Resources in Hospitality & Tourism*, vol.12, no. 2, 2013, pp. 155-174.

[19] J. J. Pigram, "Best practice environmental management and the tourism industry,"

Progress in Tourism and Hospitality Research, vol. 2, 1996, pp. 261-271.

[20] C. E. Boyd and H. R. Schmittou, "Achievement of sustainable aquaculture through environmental management," *Aquaculture Economics & Management,* vol 3, no. 1, 1999, pp. 59-69.

[21] M. N. O. Sadiku, O. D. Olaleye, H. A. Ezeka, and S. M. Musa, "Envionmental management accounting," *International Journal of Trend in Research and Development,* vol. 7, no. 5, 2020, pp. 98-100.

[22] M. Bartolomeo et al., "Environmental management accounting in Europe: Current practice and future potential," *European Accounting Review*, vol. 9, no. 1, 2000, pp. 31-52.

[23] R. D. Margerum and S. M. Born, "Integrated environmental management: Moving from theory to practice," *Journal of Environmental Planning and Management*, vol. 38, no. 3, 1995, pp. 371-392.

[24] L.Y. ShenVivian and W.Y.Tam, "Implementation of environmental management in the Hong Kong construction industry," *International Journal of Project Management,* vol. 20, no.7, October 2002, pp. 535-543.

[25] M. Chitakiraa and B. Nyikadzinoab, "Effectiveness of environmental management institutions in sustainable water resources management in the upper Pungwe River basin, Zimbabwe," *Physics and Chemistry of the Earth, Parts A/B/C,* vol. 118–119, October 2020.

[26] L. Thebeau, S. Garcia, and Richard B. Hockett, "Environmental management in contingency operations," *The Military Engineer,* vol. 101, no. 659, May-June 2009, pp. 65-66.

[27] M. Ju, L. Shi, and X. Chen, "Trends in Chinese urban environmental management," *Journal of Environmental Assessment Policy and Management*, vol. 7, no. 1, March 2005, pp. 99-124.

[28] I. Henriques and P. Sadorsky, "Environmental management practices and performance in Canada," *Canadian Public Policy / Analyse de Politiques,* vol. 39, August 2013, pp. S157- S175.

[29] M. Funder and M. Marani, "Implementing national environmental frameworks at the local level A case study from Taita Taveta County. Kenya," *Danish Institute for International Studies*, 2013.

[30] M. Watson, "Protecting the environment: The role of environmental

management systems," *The Journal of The Royal Society for the Promotion of Health,* vol. 126, no. 6, November 2006, pp. 280-284.

[31] N. Room, "Development environmental management strategies," *Business Strategy and the Environment,* vol. 1, part1, Spring 1992, pp. 11-24.

[32] S. Brammer, S. Hoejmose, and K. Marchant., "Environmental management in SMEs the UK: Practices, pressures and perceived benefits," *Business Strategy and the Environment,* vol. 21, 2012, pp. 423-434.

[33] J. Courtnell, "What is environmental management? How you can implement it today," October 2019,

https://www.process.st/environmental-management/

[34] J. A. A. Jones, *Global Hydrology: Processes, Resources and Environmental Management.* Routledge, 2014.

[35] S. Schaltegger, R. Burritt, and H. Petersen, *An Introduction to Corporate Environmental Management: Striving for Sustainability.* Routledge, 2017.

[36] W. Wehrmeyer, *Greening People: Human Resources and Environmental Management.* Routledge, 2017.

[37] J. M. Hellawell, *Biological Indicators of Freshwater Pollution and Environmental Management.* Elsevier Science Publishers, 2012.

[38] R. A. Buchholz, *Principles of Environmental Management: The Greening of Business.* Prentice Hall, 2nd edition, 1998.

[39] T. K. DaS, *Industrial Environmental Management: Engineering, Science, and Policy.* John Wiley & Sons, 2020.

[40] M. Ramkumar, K. Kumaraswamy, and R. Mohanraj (eds.), *Environmental Management of River Basin Ecosystems.* Springer, 2015.

[41] K. E. Halvorsen et al. (eds.), *A Research Agenda for Environmental Management.* Edwards Elgar Publishing, 2019.

[42] M. Russo, *Environmental Management: Readings and Cases.* SAGE Publications, 2nd edition, 2008.

[43] J. Brady, A. Ebbage, and R. Lunn, *Environmental Management in Organizations: The IEMA Handbook.* London, Routledge, 2nd Edition, 2011.

[44] D. Sarkar et al. (eds.), *An Integrated Approach to Environmental Management.* John Wiley & Sons, 2015.

[45] M. A. Camilleri, *Corporate Sustainability, Social Responsibility and Environmental Management:An Introduction to Theory and Practice with Case Studies.* Springer, 2017.

[46] I. V. M. Krishna, V. Manickam, and A. Shah, *Environmental Management: Science and Engineering for Industry.* India: BS Publications, 2017.

[47] R. Costanza, B. G. Norton, and B. D. Haskell (eds.), *Ecosystem Health: New Goals for Environmental Management.* Washington, DC: Island Press, 1992.

[48] J. Lehmann and S. Joseph, *Biochar for Environmental Management: Science, Technology and Implementation.* New York: Routledge, 2015.

[49] S. E. Curtis, *Environmental Management in Animal Agriculture.* Ames, IO: Iowa State University Press, 1983.

[50] R. V. Cooke and J. C. Doornkamp, *Geomorphology in Environmental Management: A New Introduction.* Oxford, UK: Oxford University Press, 1990.

[51] M. Keen, V. A. Brown, and R. Dyball (eds.), *Social Learning in Environmental Management: Towards a Sustainable Future.* Earthscan, 2005.

[52] M. Burgman, *Risks and Decisions for Conservation and Environmental Management.* Cambridge, UK: Cambridge University Press, 2005.

[53] T. O'Riordan (ed.), *Environmental Science for Environmental Management.* Longman, 1995.

[54] R. Welford and A. Gouldson, *Environmental Management & Business Strategy.* Pitman Publishing Limited, 1993.

[55] B. Mitchell, *Resource and Environmental Management.* Routledge, 2nd edition, 2018.

[56] C. Barrow, Environmental Management for Sustainable Development. New York: Routledge, 2nd edition, 2006.

[57] C. Allan and G. H. Stankey, *Adaptive Environmental Management.* Springer, 2009.

[58] P. R. Pryde, *Environmental Management* in the *Soviet Union.* Cambridge University Press, 1991.

[59] R. Welford (ed.), *Corporate Environmental Management 1: Systems and Strategies.* Routledge, 2nd edition, 2016.

[60] C. J. Barrow, *Environmental Management: Principles And Practice.* Routledge, 1999.

CHAPTER 10

ENVIRONMENTAL POLICY

"We need to accept the seemingly obvious fact that a toxic environment can make people sick and that no amount of medical intervention can protect us. The healthcare community must become a powerful political lobby for environmental policy and legislation." —Andrew Weil

10.1 INTRODUCTION

The relationship between environmental protection and human rights is natural. People generally support the fundamental human right to enjoy air and water free of contamination, and raise their children in an environment conducive to human life and health. Human rights may provide a mechanism for checking the global environmental excesses.

Environmental policy denotes the commitment of a government or organization to the laws, regulations, and policies on environmental issues such as air, water pollution, energy, toxic substances, and waste management. It is a commitment to communicate your organization's environmental aims and objectives to staff, customers, investors, and other external stakeholders. An environmental policy is usually a one-page written document which outlines a business' aims and principles in relation to managing the environmental impacts of a business. It sets out key aims and principles to train, educate, and inform the employees about environmental issues that may affect their work. An environmental policy does not have to exist in isolation. It may be integrated with other policies such as quality management and corporate social responsibility [1].

Environmental policy consists of two terms: environment and policy. Environment is the earth, our home— where we live, breathe, eat, and engage in other activities. Policy refers to the course of action taken by a government, party, business or individual to reach some set targets for achieving environmental goals to protect the environment and conserve its natural resources. Figure 10.1 shows different components of environmental policy [2].

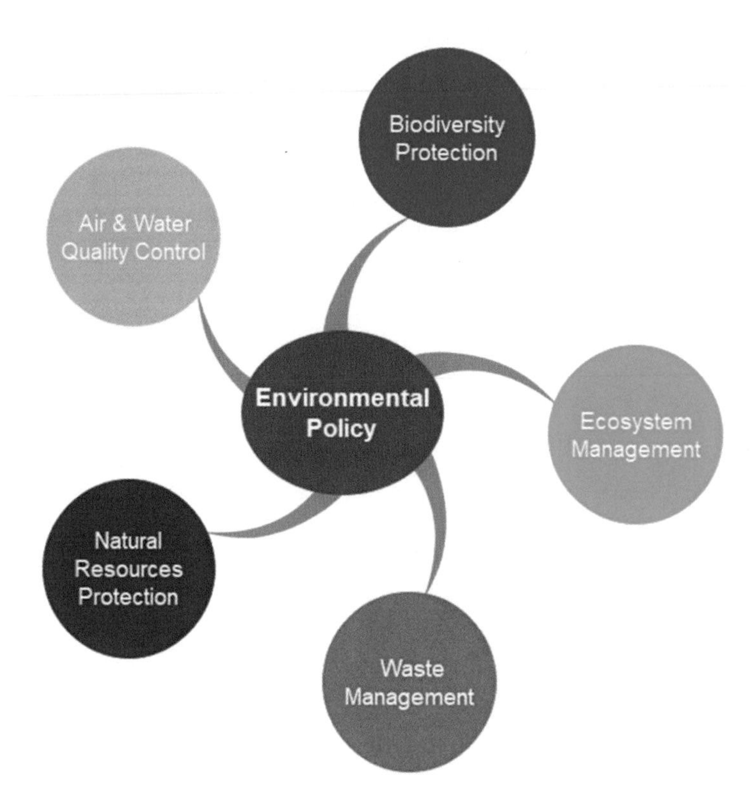

Figure 10.1 Components of environmental policy.

The standard framework in which economists evaluate environmental policies is cost–benefit analysis. This has resulted in economic-incentive or market-based policy instruments. Environmental policy efficiency can be determined through cost–benefit analysis techniques.

This chapter provides an introduction on environmental policy. It begins with explaining the concept of environmental policy. It explains policy instruments and provides some sample environmental policies. It provides some historical background on environmental policy. It presents environmental objectives to implement environmental policy. It describes duties of environmental policy professionals. It also provides different areas of applications of environmental policy. It covers how environmental policy is applied in the global environment. It highlights the benefits and challenges of environmental policy. The last section concludes with comments.

10.2 CONCEPT OF ENVIRONMENTAL POLICY

Environmental policy involves actions taken by organizations and government at the federal, state, and local level to protect the environment and conserve natural resources such as fishing, farming, and forestry. It is a commitment to the laws, regulations, and policies on environmental issues and sustainability. It may also be regarded as a mission statement for the government or organization when it comes to environmental issues. It is sometimes considered as environmental protection and conservation strategies.

Environmental policy issues include water and air pollution, chemical and oil pills, smog, land conservation and management, and wildlife protection. Environmental policy making is no longer the exclusive domain of government; private or public organizations assume responsibility for the environment. International environmental policies and agreements are increasingly important in a globalized economy. Environmental issues are now legitimate concerns for many companies and organizations such as the World Bank, IMF, and the WTO.

In his speech "Seven Principles of Sound Public Policy," Larry Reed brought together a list of foundational truths to help guide the creation of effective public policy. The following seven basic principles (summarized here) will provide legislators and regulators with a solid foundation for establishing policies that can effectively manage our beautiful and productive natural areas [3].

1. Environmental stewardship starts with individuals, not politicians or bureaucracies.
2. Property rights are the most basic of human rights and an essential foundation for environmental stewardship.
3. Competition and voluntary cooperation fosters innovation and wise use of natural resources.
4. Efficiency is the key to reducing environmental impacts.
5. Harming properties harms the environment.
6. Top-down approaches rarely work.
7. Technological innovation is the key to improving the environment.

The rise of a global environmental movement is perhaps one of the most significant efforts for protecting the global environment, especially in developing countries. The number of environmental nongovernmental organizations addressing international issues has exploded. Nearly every nation has at least one environmental organization that collaborates with their colleagues from other nations. Leading environmental activists such as Chico Mendes and Ken Saro Wiwa have been killed [4]. A typical display of environmental policy is illustrated in Figure 10.2 [5], which shows that Amtek, manufacturer of various connectors, is committed to green policy: energy conservation, waste reduction, resource recycling, and regulatory compliance.

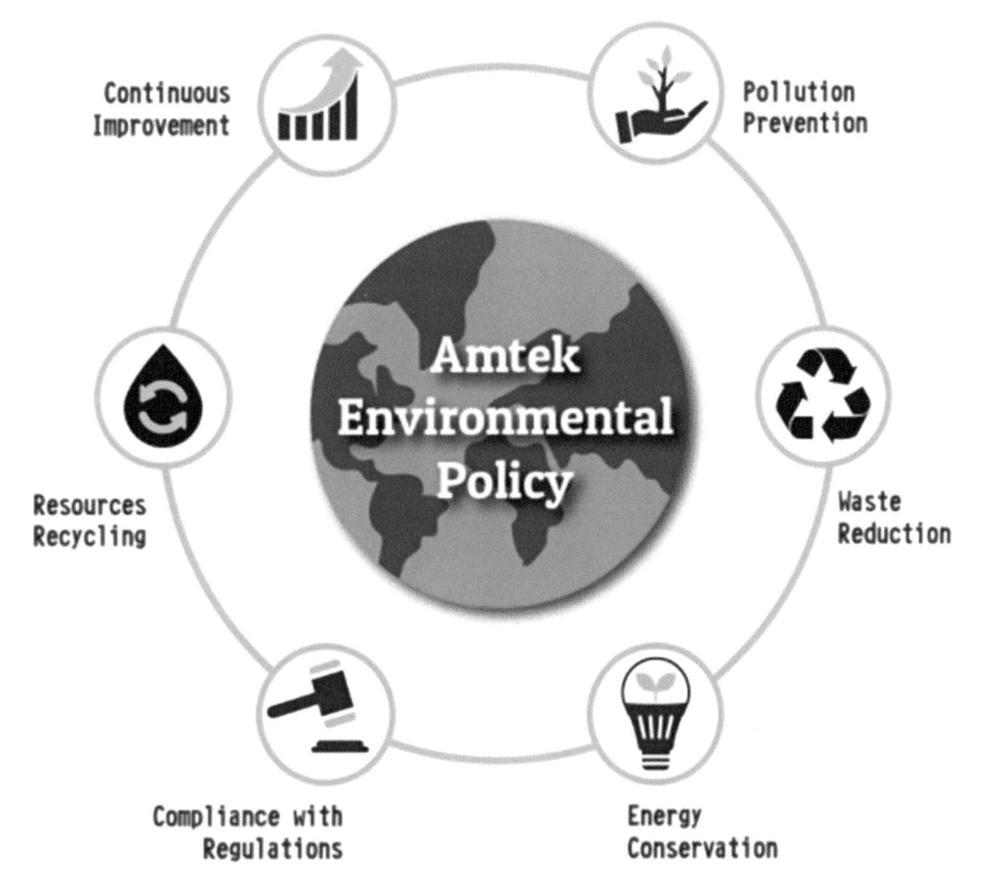

Figure 10.2 A typical display of environmental policy [5].

10.3 POLICY INSTRUMENTS

Environmental policies usually combine the identification of a goal with some means to achieve that goal, known as "instruments." Environmental policy instruments are essentially different tools used by governments or organizations to implement their environmental policies. Common environmental policy instruments are [6]: (1) Economic Incentives and Market-based Instruments, (2) Voluntary Agreements, (3) Regulatory Instruments, (4) Mixed Instruments, and (5) International Framework.

The choice of policy instrument is a crucial environmental policy decision. There is no universally right choice of instrument for managing a nation's environment. All policy instruments require monitoring capability, enforcement resolution, and control of corruption. Common instruments include emissions taxes, tradable emissions allowances, government subsidy reductions for emissions, performance standards, and mandates for the adoption of specific technologies. No single instrument is clearly superior along all the dimensions relevant to policy choice. It is sometimes desirable to design hybrid instruments that combine various features [7].

10.4 SAMPLE ENVIRONMENTAL POLICIES

Although there is no standard format or content for writing an environmental policy, it is important that it is done carefully. We present the following three examples of environmental policy [8,9]:

Example 1: Since 1974, ABC Farms has operated a farrow to finish hog farm in ABC County. We at ABC Farms try to produce a consumer-safe product by following quality assurance practices that emphasize good management in the handling and use of animal health products and review of our approach to herd health programs. We strive to protect the environment by minimizing waste within our control and following best management practices. ABC Farms is committed to complying with regulations and is committed to continual improvement of its management practices.

Example 2: ABC Farms Inc. is committed to meeting or exceeding relevant environmental regulations and other environmental related requirements through the continual improvement of its environmental management system and the prevention of pollution. ABC Farms will develop and monitor annual environmental objectives and targets to assist in meeting this commitment.

Example 3: (Company Name) is committed to reducing its impact on the environment. We will strive to improve our environmental performance over time and to initiate additional projects and activities that will further reduce our impacts on the environment. Our commitment to the environment extends to our customers, our staff, and the community in which we operate. We are committed to:

- Comply with all applicable environmental regulations;
- Prevent pollution whenever possible;
- Train all of our staff on our environmental program and empower them to contribute and participate;
- Communicate our environmental commitment and efforts to our customers, staff, and our community; and
- Continually improve over time by striving to measure our environmental impacts and by setting goals to reduce these impacts each year.

Signed: ______________________________

Date: ______________________________

Printed/Title: ______________________________

10.5 HISTORICAL BACKGROUND

Written policies aimed at environmental protection date back to ancient times. The United Nations founded its United Nations Environmental Program (UNEP) as early as 1972. Since then, some environmental agreements (such as Montreal Protocol, the Kyoto Protocol, the Convention on Global Climate Change, and the Convention on Biodiversity), have risen under the umbrella of the United Nations. The Kyoto Protocol is the single most important instrument of international climate policy to date. This Protocol was the first legally binding international commitment by industrialized countries to reduce their emissions by at least 5 percent by 2012 compared to 1990. Also since the early 1970s, environmental policies have shifted from end-of-pipe solutions to prevention and control. The first generation of environmental policy was marked by ungainly bureaucracies, high costs, political polarization, and a litigious atmosphere.

Several instruments have been developed to influence the behavior of those who contribute to environmental problems. These include regulation, financial incentives, and environmental reporting and ecolabeling. A guiding principle is the "polluter pays" principle, which makes polluters liable for the costs of environmental damage, and the precautionary principle [10].

Today's communication networks such as the internet has also increased the effectiveness of the global environmental community considerably. The Internet helps maintain global networks, share information, and coordinate international efforts.

10.6 ENVIRONMENTAL OBJECTIVES

An organization may set environmental objectives which are regularly reviewed, to ensure that their actions effectively implement their environmental policy. The following are typical objectives of an organization [11]:

- To take significant environmental aspects and impacts into account throughout our operations, maintaining a functioning environmental management system at each factory.
- To ensure that environmental issues are properly assessed and considered when key decisions are taken about supply chains, processes and new product development.
- To establish and measure the significant environmental impacts of our operations, set targets for performance improvements and monitor progress against those targets in areas including but not limited to energy, greenhouse gas emissions, water usage / quality and waste.
- To use energy and natural resources wisely and efficiently, eliminate and minimize waste, and re-use and recycle where practical.
- To make a real and meaningful contribution to mitigating climate change and global water scarcity, by reducing greenhouse gas emissions and water impact across the complete lifecycle of our products and their packaging, reflecting national and international government agendas when setting targets.
- To engage with our suppliers, customers, and other stakeholders on environmental issues, including the sustainability of our raw materials, and packaging material supply chains.
- To ensure that employees have a level of knowledge and understanding appropriate to their environmental responsibilities and are aware of actions they can take to reduce their impacts.
- To conduct an annual review, including progress against targets, and to make that review publicly available in our annual Sustainability Report.

10.7 ENVIRONMENTAL POLICY PROFESSIONALS

Typically, careers in environmental policy require a Masters degree. Most careers in environmental policy involves planning, and collaborating with co-workers, overseeing building programs, and spending time outdoors. Environmental policy professionals should be familiar with the following tasks [12]:

- Evaluate land use issues
- Remain compliant with applicable zoning and engineering plans
- Organize correspondence, reports, data, and other project information
- Research local, state and federal environmental policies
- Analyze remediation strategies
- Proactively maintain client and stakeholder relationships
- Show innovation in strategy across various projects
- Work seamlessly in an interdisciplinary environment with other technicians, scientists, and engineers
- Create and maintain technical documentation
- Collect and analyze data
- Maintain records and databases to support standard project operations
- Provide technical feedback on environmental policy
- Maintain aspects of operating budgets
- Commit to best practices in work and research

10.8 APPLICATIONS OF ENVIRONMENTAL POLICY

Environmental policy is commonly applied in the following areas [13,14].

- *Environmental Regulation:* This emerges from an analytical exercise involving the maximization of social welfare. Governments often control the environmental conduct of businesses under their jurisdiction by imposing and enforcing environmental regulations. Environmentalists, businesses, organizations, and other interest groups interact to develop environmental legislation. For example, the adoption of pollution-reducing technology calls for regulation. The atmosphere for environmental regulation has changed dramatically. The success of international agreements depends on the capacity of national governments to impose environmental regulations, and ensure that they endure over time. In the US, environmental regulation is frequently implemented through a system that allows states to choose whether to assume primary authority for implementation and enforcement.
- *Environmental Interest Groups:* The issue of interest groups is a complicated one in environmental policy. The interest (or lobby) groups, which may take the form of lobby groups, contributes money in order to influence policy decisions. Sometimes, we find two opposing groups: environmental advocacy organizations in opposition to trade associations representing business interests. Also, there may exist a substantial number of groups with a stake in the choice of policy instruments and their level of stringency: environmental organizations, business interests, labor unions, administrative and trial lawyers, government agencies themselves, as well as the general public. The final choice of regulatory instruments will be an outcome of a process of interaction between policy-makers and the various interest groups that bring pressure to bear on these decisions.

- *Political Economy:* Economics plays a vital role in the design of new policy measures. Economists and economic instruments have had a modest impact on shaping environmental, health, and safety regulation. The predictions of environmental policies at present are based mainly on mainstream economics and the standard framework in which economists evaluate environmental policies is cost–benefit analysis. The political economy of environmental policy is impressive and promising. The actual practice of environmental decision-making has been centralized. One may consider a system of environmental policy-making in which the central government sets standards and oversees measures to address explicitly national pollution problems. Under myopic environmental policies, emission taxes tend to provide a stronger incentive to invest in both R&D and adoption of new technology as compared to emission allowances.
- *Adoption of Technology:* Technological change can help the economic models of environmental policy. Environmental policy interventions themselves create new constraints and incentives that affect the process of technological developments. Adoption of new technology can help environmental policy. The processes of technological change are characterized by a substantial uncertainty associated with the time of arrival and performance of new technologies. Companies may decide whether or not to adopt the new technology. Some may study innovation and R&D before they adopt the new technology.

Other areas in which environmental policy is being applied include public health, transportation, air quality, environmental problems in agriculture, risk reductions in energy generation, cross-national policy convergence, green public procurement, and natural gas use.

10.9 INTERNATIONAL ENVIRONMENTAL POLICY

Protecting the global environment has emerged as one of the major challenges in the international community. Some developing countries have established laws and policies to address environmental problems. International environmental policy is important in times of increasing globalization. Several global environmental issues (such as global warming and biodiversity, spill over national borders) cannot be resolved by a single nation acting alone [15]. They can only be properly addressed through international cooperation and a form of international agreement. International environmental policy covers a number of issues: climate change, the need for decarbonization, sustainable energy policy, the loss of biodiversity, desertification, sustainable waste management, protection against hazardous substances, and the conservation of land, forests, and seas. A list of international environmental agreements can be found in [16].

The United Nations founded its United Nations Environmental Program (UNEP) as early as 1972. Since then, the number of international environmental policy has increased steadily. The environment is included in the agenda of the G8 (an inter-governmental political forum), which comprises of Germany, Japan, Great Britain, United States of America, France, Italy, Canada, and Russia.

We now consider how environmental policy is implemented in some countries.

- *United States:* Environmental policy in the US involves governmental actions at the federal, state, and local level to protect the environment and conserve its natural resources. It has always been characterized by political conflict. The US National Environmental Policy Act (NEPA) established the broad national framework for protecting the environment. NEPA assures that all branches of government give proper consideration to the environment before undertaking any federal action that affects the environment. It requires federal agencies to assess the environmental effects of their proposed actions before making decisions. Its requirements are invoked when airports, buildings, military complexes, highways, parkland purchases, and other federal activities are proposed [17]. NEPA's decision process is illustrated in Figure 10.3 [18]. Although the US is the world's only remaining economic and political superpower, it is also the largest polluter and the largest consumption of natural resources. The US has always opposed the development of any general environmental covenant, fearing that the resulting principles would be weak.

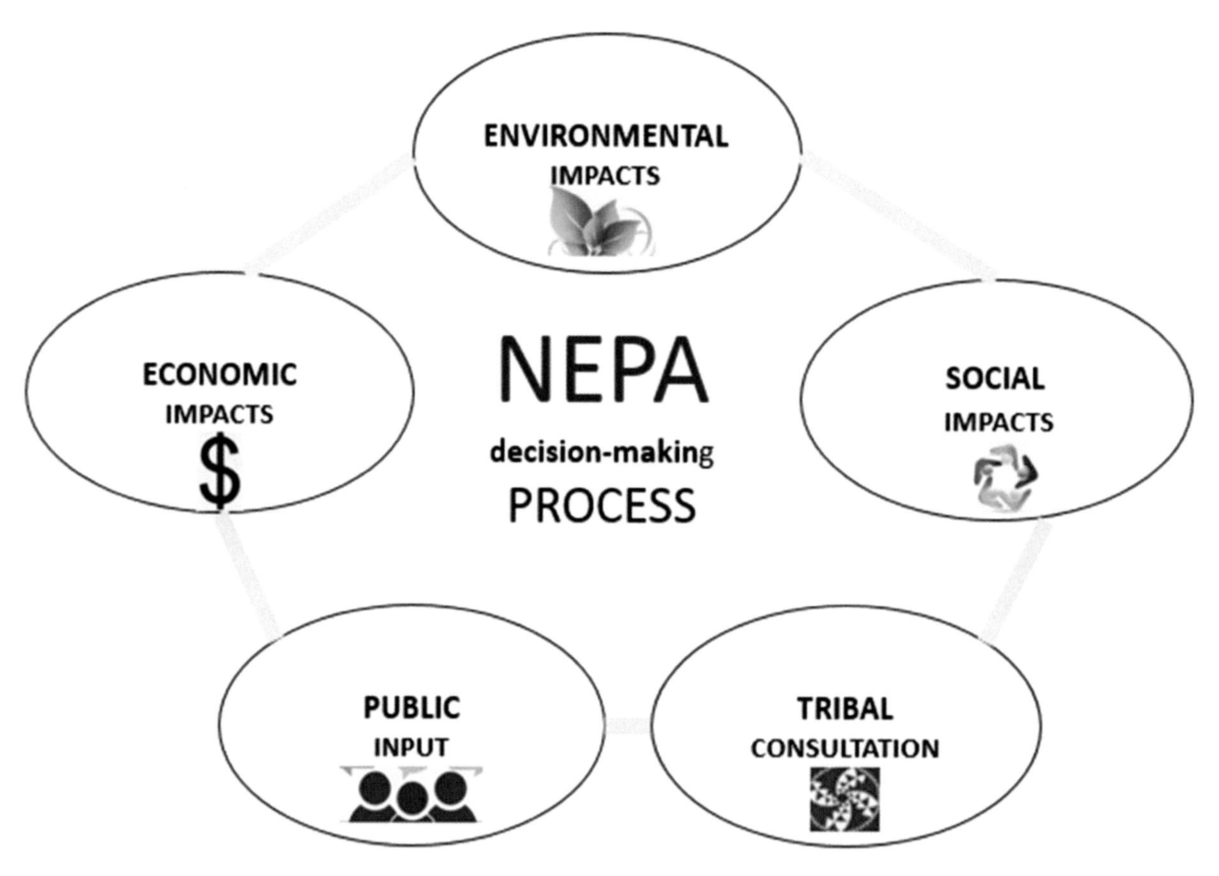

Figure 10.3 NEPA's decision-making process[18].

- *European Union:* The European Union is active and setting the pace in international environment policy. European citizens enjoy some of the world's highest environmental standards. It is a set of strategies, actions, and programs to promote climate resilient society and thriving economy in its natural environment. Some EU nations have long environmental protection traditions, which are also reflected in their environmental law. The European environmental research and innovation policy aims at implementing a transformative agenda to greening the economy and the society so as to achieve a truly sustainable development. Political support of the principle of environmental policy integration (EPI) is particularly strong in the European Union. Figure 10.4 depicts the influence on European Union's environmental policy by nationality [19]. EU leadership in international environmental politics is best explained "regulatory politics," which combines the effects of domestic politics and international regulatory competition. Several factors have contributed to the rise of the EU as the undisputed leader in global environmental governance [20].

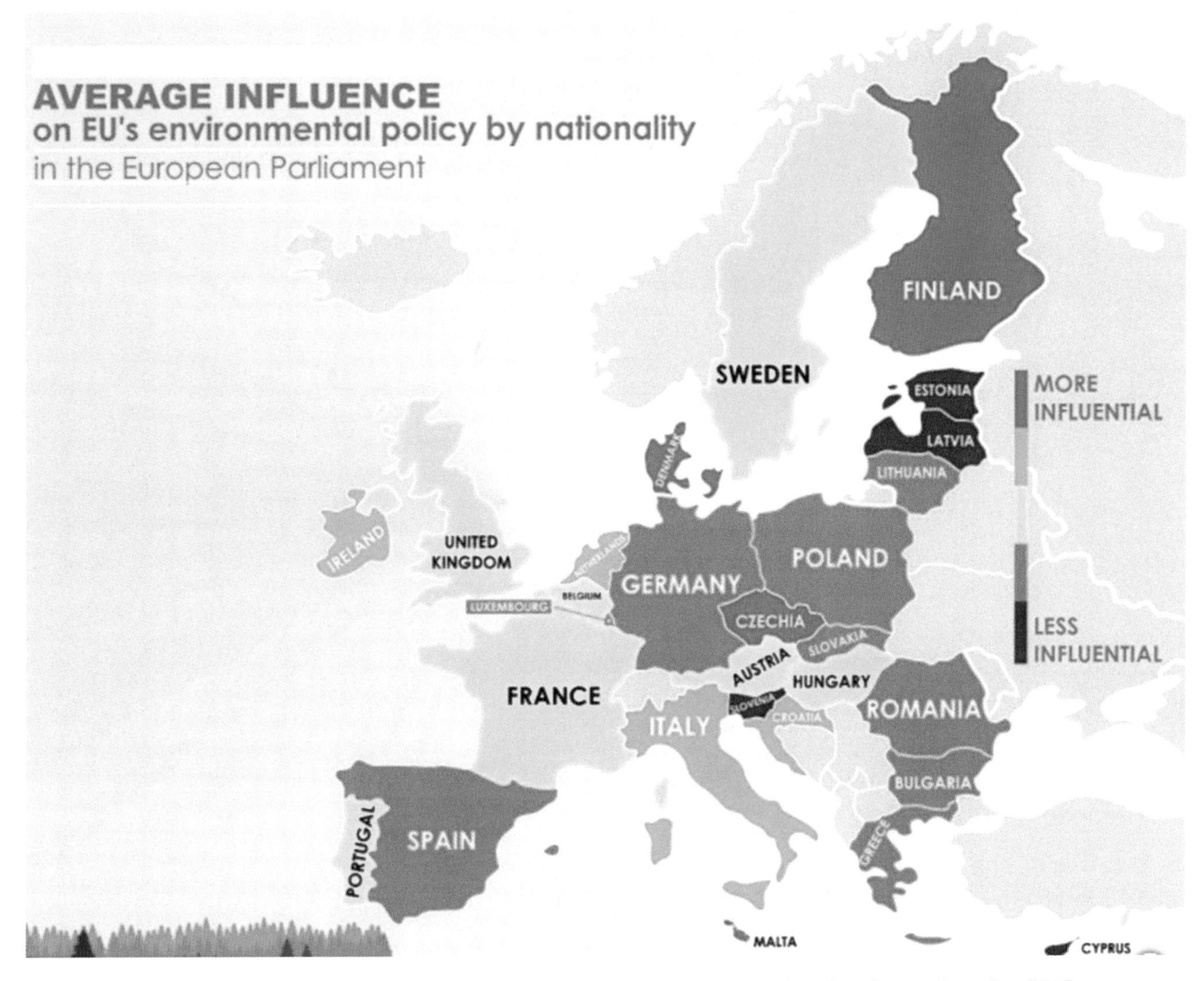

Figure 10.4 Influence on European Union's environmental policy by nationality [19].

- *Finland:* Environmental rights and responsibilities have a long history in Finland. It is a nation that has prided itself on its nature-centric culture. This nation is one of few industrialized democracies that consistently performs well in future-oriented environmental policy. This is due to strategies, which penetrate the country's governance structure. Since entering the EU in 1995, Finland has been considered as one of the "forerunners" with regard to environmental policy. Finland demonstrates an example where commitment to long-term perspectives has become embedded in institutions that produce environmental policy [21].
- *Russia:* Russia and the other parts of the Soviet Union never really had rule-of-law traditions. The hard realities of environmental compliance were basically unknown to industry in the Soviet bloc nations. Monitoring in the former Soviet bloc usually measured ambient air quality, not what pollutants plants released at the end of their discharge pipes. The former Soviet bloc had considerable economic, cultural, and political baggage that was directly relevant to the introduction of any state-of-the-art environmental tools, including market-based instruments.
- *China*: With its rapid economic growth in recent decades, China is generating growing concern about its environmental record. Serious environmental problems, including deforestation, erosion, and pollution of air, land, and water have prevailed throughout China. These have resulted in changing patterns of resource use. An environmental policy in China introduced

stricter regulations on sulfur dioxide (SO_2) emissions in targeted cities. Although China has strong environmental protection laws and policies, they are not being effectively implemented in some rural areas, where industrialization has resulted in increased financial well-being as well as contributed to decreased environmental quality [22].

- *Sri Lanka*: This island has a population of 19 million, placing it among the most densely populated nation in the world. Sri Lanka's constitution states that: "The duty of every person in Sri Lanka is to protect nature and conserve its riches." An appreciation of the role of civil society in maintaining a healthy environment is lacking in Sri Lanka [23].
- *United Kingdom:* Since the early 1990s there has been increased environmental legislative pressure and growing public awareness of environmental issues. Environmental policies have not been successful in encouraging more environmentally proactive behavior within small and medium-sized enterprises (SMEs) in the UK. The poor environmental performance is due to a lack of institutional enfranchisement for SMEs. Companies in UK endeavor to comply with increasing environmental regulation and have no problem adopting ISO 9000 [24]. In UK, local government and a network of environmental organizations develop environmental policy to control and manage the local impacts on the environment. The use of cost-benefit analysis (CBA) in British environmental policy has gone through some stages. CBA has started to have influence in the use of environmental taxes.
- *Tanzania*: Tanzanians are intimately connected to the environment: their survival and that of their future generations depends on the harmonious relationship with the natural elements. They do not have the dispensable luxury of ignoring the fundamental stresses at the interface of development and environment. They somehow believe that a healthy economy and a healthy environment go hand-in hand. There is a clear cause-and-effect relationship between poverty and environmental degradation. Satisfaction of basic necessities is, therefore, an environmental concern of relevance to environmental policy. The government of Tanzania is committed to sustainable development through short, medium, and long term social and economic growth [25].
- *Botswana:* Environmental degradation in Botswana is rooted in structural problems, such as cultural continuity in rural-urban settings, population growth and concentration of population in a few urban and settlements, and structural poverty. As the environment's potential declined, however, the people continued to practice their traditional modes of utilization without realizing that change was required. Structural poverty is also the inability of the person to meet his/her basic needs due to constraints created by social structures. The current environmental policy of the Republic of Botswana pays most attention to the utilization of the land and the needs of livestock. It tends to emphasize technical solutions to environmental problems and issues, while ignoring cultural issues in the formulation of policy [26].

10.9 BENEFITS

Environmental policy is concerned with how best to govern the relationship between humans and the natural environment. It is probably the only policy domain with the main goal of protecting nonhuman entities. There are some benefits of having an environmental policy. A sound environmental policy can help you manage your resources, achieve cost savings, and promote more environmentally friendly practices. A good environmental policy can help businesses operate in a responsible, environmentally sound and sustainable manner, manage their resources, achieve cost savings, and promote more environmentally friendly practices.

Environmental policy can also provide the following significant benefits to an organization [27]:

- helping you to stay within the law
- improving information for employees about their environmental roles and responsibilities
- improving cost control
- reducing incidents that result in liability
- conserving raw materials and energy
- improving your monitoring of environmental impacts
- improving the efficiency of your processes

The benefits are not restricted simply to internal operations. By demonstrating commitment to environmental management, one can develop positive relations with external stakeholders, such as investors, insurers, customers, suppliers, regulators, and the local community. International environmental agreements are becoming indispensable in a globalized economy.

10.10 CHALLENGES

There are some challenges or limitations environmental policies have to overcome. We need to address fundamental questions on who should determine our environmental goals or objectives. We also must figure out what the next generation of environmental policy will be. Non-point source pollutants pose an important challenge to the continuing campaign to clean the environment. Traditional command-and-control regulations work poorly against the pollution problems caused by non-point sources, which are virtually universal. Another challenge is the growing recognition of pollution problems such as global warming that knows no national boundaries. This requires collective-action and policy strategies that encompass the world. The challenge of the second-generation environmental policy is to do everything that the first generation of environmental policy did and do it better [28].

Although international environmental interdependence is growing, the international system lacks a central authority to foster environmental protection. Increased environmental concern leads to a higher dirt tax, a lower tax on labor, less employment, more public abatement, and a cleaner environment. Some environmentalists are concerned that trade liberalization may damage the environment by giving governments incentives to relax environmental policies so that domestic producers may remain competitive. In spite of the growing international environmental interdependence and the globalized economy, the international community lacks a central authority to foster environmental protection. Many environmentalists are of the view that pollution taxes or systems of tradeable emission permits entail "putting the environment up for sale" and are consequently immoral and unacceptable.

10.11 CONCLUSION

Careful stewardship of the planet is very important and should not be ignored. Environmental policy is a goal-oriented plan or measure by the government or organization for the stewardship of the earth's natural resources regarding the effects of human activities on the environment, particularly those measures that are designed to protect and conserve the environment from the harmful effects of human activities on the ecosystems. The main goal of environmental policy is to regulate resource use or reduce pollution (such as water and air pollution, chemical and oil spills, and smog) in order to promote human welfare and/or protect natural systems. The modern fusion of business and environmental qualifications means that there is a growing demand for leaders with advanced degree in environmental science. Some academic institutions now offer Master's degree on environmental policy. More information about environmental science can be found in books in [29-74] and related journals:

- *Environment Policy*
- *Environmental Politics*
- *Journal of Environmental Policy and Planning*
- *Environmental Policy and Law*
- *Environmental policy and governance*
- *Issues in Energy and Environmental Policy*
- *Review of Environmental Economics and Policy*

REFERENCES

[1] M. N. O. Sadiku1, U. C. Chukwu, A. Ajayi-Majebi, and S. M. Musa, "Essence of environmental policy," *Journal of Scientific and Engineering Research*, vol. 7, no. 11, 2020, pp. 156-164.

[2] "Diagram of environmental policy,"

https://www.canstockphoto.com/diagram-of-environmental-policy-33270235.html

[3] J. Hayes & T. Myers, "Seven principles of sound environmental policy,"

https://www.mackinac.org/27385

[4] D. Hunter, "Global environmental protection in the 21st century," June 1999,

https://ips-dc.org/global_environmental_protection_in_the_21st_century/
[5] "Environmental policy,"

https://amtek-co.com.tw/environmental-policy/

[6] J. Islam, L. Ferdous, and A. Begum, "The environmental policies and issues to ensure safe environment," *International Journal of Innovative Research in Engineering & Science,* vol. 10, no. 3, October 2014.

[7] L. H. Goulder and I. W. H. Parry, "Instrument choice in environmental policy,"

Review of Environmental Economics and Policy, vol. 2, no. 2, Summer 2008, pp. 152–174.

[8] "Environmental policy examples,"

http://www.p2pays.org/porktool/samples/Policy.pdf

[9] "Sample environmental policy statement,"

https://mde.state.md.us/marylandgreen/Documents/mdgreen/MGR_Policy.pdf

[10] E. van Bueren, "Environmental policy,"

https://www.britannica.com/topic/environmental-policy

[11] R. Kapoor, "Environment policy,"

https://www.rb.com/media/782/english-environmental-policy.pdf

[12] "Environmental policy & planning careers,"

https://www.environmentalscience.org/careers/environmental-policy-and-planning

[13] W. E. Oates and P. R. Portney, "Chapter 8 - The Political Economy of Environmental Policy," *Handbook of Environmental Economics*, vol. 1, 2003, pp. 325-354.

[14] A. Löschel, Technological change in economic models of environmental policy: A survey," *Ecological Economics*, vol. 43, no. 2-3, December 2002, pp. 105-126.

[15] Bell, R. Greenspan, and C. Russell. "Environmental policy for developing countries." *Issues in Science and Technology*, vol. 18, no. 3, Spring 2002.

[16] "List of international environmental agreements," *Wikipedia*, the free encyclopedia.

https://en.wikipedia.org/wiki/List_of_international_environmental_agreements

[17] "Summary of the national environmental policy act,"

https://www.epa.gov/laws-regulations/summary-national-environmental-policy-act

[18] "EnviroProjects can manage your entire NEPA documentation process or contribute to topic-specific elements of your NEPA assignments,"

http://www.enviroproj.com/services/nepa-compliance-services/

[19] "Tints of green: Who influences environmental policy tints of green: who influences environmental policy in the European parliament and how?"

https://www.votewatch.eu/blog/tints-of-green-who-influences-environmental-policy-in-the-european-parliament-and-how/arliament and How

[20] R. D. Kelemen, "Globalizing European Union environmental policy," *Journal of European Public Policy,* vol.17, no. 3, 2010, pp. 335-349.

[21] V. Koskimaa, and L. Rapeli, and J. Hiedanpää, "Governing through strategies: How does Finland sustain a future-oriented environmental policy for the long term?" *Futures,* vol. 125, January 2021.

[22] K. E. Swanson, R. G. Kuhn, and W. Xu, "Environmental policy implementation in rural China: A case study of Yuhang, Zhejiang," *Environmental Managemen*t, vol. 27, 2001, pp. 481–491.

[23] "National environmental policy and strategies,"

https://www.preventionweb.net/files/15417_nationalenvironmentpolicyandstrateg.pdf

[24] A. Revell and R. Rutherfoord, "UK environmental policy and the small firm: broadening the focus," *Business Strategy and the Environment,* vol. 12, no. 1,

January/February 2003, pp. 26-35.

[25] The United Republic of Tanzania, "National environmental policy," December 1997.

[26] M. H. Abucar and P. Molutsi, "Environmental Policy in Botswana: A Critique," *Africa Today*, vol. 40, no. 1, 1st Qtr., 1993, pp. 61-73.

[27] "How to write an environmental policy,"

https://www.infoentrepreneurs.org/en/guides/how-to-write-an-environmental-policy/

[28] D. F. Kettl, "Environmental policy: The next generation," October 1998,

https://www.brookings.edu/research/environmental-policy-the-next-generation/

[29] A. Burke, *Development and Environmental Policy in India.* Createspace Independent Publishing. 2018.

[30] C. Knill and D. Liefferink, *Environmental Politics in the European Union: Policy-Making, Implementation and Patterns of Multi-Level Governance.* Manchester University Press, 2007.

[31] A. I. Herson and G. A. Lucks, *California Environmental Law & Policy*. Solano Press Books; 2nd edition, 2017.

[32] R. Kemp, *Environmental Policy and Technical Change: A Comparison of the Technological Impact of Policy Instruments.* Edward Elgar Pub., 1997.

[33] L.K. Caldwell, *International Environmental Policy: Emergence and Dimensions.* Durham, NC: Duke University Press, 1990.

[34] W. J. Baumol and W. E. Oates, *Economics, Environmental Policy, and the Quality of Life.* Aldershot, UK: Gregg Revivals, 1993.

[35] C. Carraro, Y. Katsoulacos, and A. Xepapadeas, *Environmental Policy and Market Structure.* Springer Science & Business Media, 2013.

[36] M. S. Andersen and D. Liefferink, *European Environmental Policy: The Pioneers.*

Manchester University Press, 1999.

[37] C. Knill and A. Lenschow, *Implementing EU Environmental Policy: New Directions and Old Problems.* Manchester University Press, 2000.

[38] M. Nilsson and K. Eckerberg (eds.), *Environmental Policy Integration in Practice: Shaping Institutions For Learning.* Taylor & Francis, 2009

[39] H Weidner and M Jänicke (eds.), *Capacity Building in National Environmental Policy: A Comparative Study of 17 Countries.* Berlin, Germany: Springer Verlag, 2002.

[40] A. Bento and I. Parry, *Tax Deductions, Environmental Policy, and the Double Dividend Hypothesis.* The World Bank, 1999.

[41] L. K. Caldwell and P. S. Weiland, *International Environmental Policy: From the Twentieth to the Twenty-First Century.* Duke University Press, 3rd edition, 1996.

[42] F. R. Anderson, *NEPA in the Courts: A Legal Analysis of the National Environmental Policy Act.* Taylor & Francis, 2013.

[43] H. Buller, G. A. Wilson, and A. Holl, *Agri-Environmental Policy in the European Union.* Ashgate Publishing Ltd., 2000.

[44] B. J. Sinkule and L. Ortolano, *Implementing Environmental Policy in China.* Westport, CT: Praeger Publishing, 1995.

[45] A. Jordan and A. Lenschow (eds.), *Innovation in Environmental Policy? Integrating the Environment for Sustainability.* Edward Eglar Publishing, 2020.

[47] N. Haigh, *EEC Environmental Policy and Britain.* Longman, 1987.

[48] D. A. Mazmanian and M. E. Kraft (eds.), *Toward Sustainable Communities: Transition and Transformations in Environmental Policy.* Cambridge, MA: MIT Press, 2009.

[49] D. Wallace, *Environmental Policy and Industrial Innovation: Strategies in Europe, the USA and Japan.* Routledge, 1995.

[50] R. B. Mitchell, *Intentional Oil Pollution at Sea: Environmental Policy and Treaty Compliance.* Cambridge, MA: MIT Press, 1994.

[51] R. Therivel and B. F. D. Barrett, *Environmental Policy and Impact Assessment in Japan.* Routledge, 2020.

[52] E. Ray Clark and Larry W. Canter (eds.), *Environmental Policy and NEPA: Past, Present, and Future.* Boca Raton, FL: CRC Press, 1997.

[53] C. Adelle, K. Biedenkopf, and D. Torney (eds.), *European Union External Environmental Policy: Rules, Regulation and Governance Beyond Borders.* Palmgrave Macmillan, 2018.

[54] S. Cohen, *Understanding Environmental Policy.* Columbia University Press, 2006.

[55] M. E. Kraft and Kamieniecki (eds.), The Oxford Handbook of U. S. *Environmental Policy.* Oxford University Press, 2013.

[56] T. Sterner (ed.), *The Economics of Environmental Policy: Behavioral and Political Dimensions.* Edward Elgar Publishing, 2016.

[57] D. M. Konisky (ed.), *Handbook of U.S. Environmental Policy.* Edward Elgar Publishing, 2020.

[58] W. J. Baumol and W. E. Oates, *The Theory of Environmental Policy.* Cambridge University Press, 2nd edition, 2012.

[59] R. N. L. Andrews, *Managing The Environment, Managing Ourselves: A History Of American Environmental Policy.* Yale University Press, 2nd edition, 2006.

[60] A. Jordan, *Environmental Policy in the European Union.* Routledge, 2005.

[61] A. Jordan and C. Adelle, *Environmental Policy in the EU: Actors, Institutions and Process.* Routledge, 3rd edition, 2013.

[62] T. Delreux and S. Happaerts, *Environmental Policy and Politics in the European Union.* Red Globe Press, 2016.

[63] M. R. Greenberg, *Environmental Policy Analysis and Practice.* Rutgers University Press, 2008.

[64] S. Rinfret and M. C. Pautz, *US Environmental Policy in Action.* Palmgrave Macmillan, 2019.

[65] P. Leroy and A. Crabb, *The Handbook .oOf Environmental Policy Evaluation.* Earthscan, 2012.

[66] A. Xepapadeas, *Advanced Principles in Environmental Policy.* Edward Elgar, 1997.

[67] C. H. Eccleston and F. March, *Global Environmental Policy: Concepts, Principles, and Practice.* Boca Raton, FL: CRC Press, 2011.

[68] N. J. Vig and M. E. Kraft, *Environmental Policy: New Directions for the Twenty-First Century.* Sage Publicatons, 10th edition, 2019.

[69] M. Hessing and T. Summerville, *Canadian Natural Resource and Environmental Policy: Political Economy and Public Policy.* UBC Press, 2nd edition, 2014.

[70] C. Carraro and F. Lévêque, *Voluntary Approaches in Environmental Policy: An Assessment.* OECD Publishing, 2013.

[71] M. S. Bhatt, S. Ashraf, and A. Illiyan (eds.), *Problems and Prospects of Environment Policy: Indian Perspective.* Aakar Books, 2008.

[72] H. Addams, J. L. R. Proops (eds.), *Social Discourse and Environmental Policy: An Application of Q Methodology.* Edward Elgar, 2000.

[73] M. E. Kraft, *Environmental Policy And Politics.* Routledge, 6th edition, 2017

[74] J. van Tatenhove, B. Arts and P Leroy (efs.), *Political Modernisation and The Environment: The Renewal of Environmental Policy Arrangements.* Springer, 2013.

CHAPTER 11

ENVIRONMENTAL LAW

"Environmental laws give power to the people. Republicans can huff, puff and scream about what they consider strict regulations, but when they cry out for reform, for a quicker process, they're really calling for a restriction of the rights of people to be involved in the planning process." - Anonymous

11.1 INTRODUCTION

The world is getting into a smaller place and becoming a single environment. Globalization has increased the economic interconnection of nations, enabling the free flow of goods, capital, people, and pollution across borders. Concern for the environment is embedded in the major religious traditions. Environmental problems now transcend national borders and pose serious challenges to the health of the planet. The impact of humanity on the planet has become so profound that scientists believe we have entered an epoch known as the Anthropocene. The word "Anthropocene" refers to the period in which people have a devastating and overwhelming impact on the earth and its systems. In the Anthropocene, humans as ecological agents have changed and continue to change the earth and its natural system [1].

Laws are put in place for several reasons. Some laws are to curb human excessiveness and selfishness. Some are for the public good, health, and protection. Others are designed to avoid some forms of irreparable harm. Law may also serve as a means of exerting power and control. Governments use laws and other tools to protect our economy, health, and environment. Protecting the environment usually involves active employment of capital, labor, and other scarce resources which could have been used for other purposes [2].

The environment is a public good that needs to be protected. Humans are directly dependent on their environment and natural resources. They also rely on a stable economy, which is fueled by natural resources. Industrial and other economic activities have significant impact on environment. Environmental law is the law that applies to environmental problems. It helps to ensure the environment and the economy are equally protected and promoted. It is essentially the collection of laws, regulations, and agreements that governs how humans interact with the natural environment. It includes both the regulation of pollutants, biodiversity and natural resource conservation, energy, agriculture, real estate, waste management, and land use. It embraces green initiatives, sustainability strategies, and alternative energy sources. It has expanded to include international environmental agreements, international trade, and environmental justice [3].

This chapter provides an introduction on environmental law in the United States and other nations around the world. It begins by discussing what environmental law is all about. It addresses the duties of the environmental lawyers. Then it presents the characteristics, principles, and scope of environmental law. It covers international environmental law, its scope, and principles. It provides some applications of environmental law. It addresses environmental law in developed nations as well as developing nations. It highlights the benefits and challenges of environmental law. The final section concludes with comments.

11.2 CONCEPT OF ENVIRONMENTAL LAW

Environmental laws (also known as environmental and natural resources laws) are the laws that deal with environmental challenges. They are legal enactments designed to consciously preserve the environment or protect the environment from damage. They are meant to protect human health as well as the environment. The laws cover pollutants, natural resource conservation, energy, farming, and land use. The laws also provide a guideline so that we can take care of the environment in an effective manner

Environmental laws come from a variety of sources. It is basically a combination of federal, state and local laws, regulations, policy choices, science, and health concerns. For most countries, the national or federal constitutions of the nation are the primary sources of environmental laws. In the US, most federal regulations come from the Environmental Protection Agency (EPA). State and federal environmental laws are the basis for the development of regulations designed to protect the air, water, and food supply, and to control pollutant discharge. Thus, the sources of environmental law can be summarized as follows [4]:

- *International Law:* This is the body of rules established by custom or treaty and recognized by nations as binding in their relations with one another.
- *Common Law:* This consists of custom and judicial precedent rather than statute.
- *Statutory Law*: These are written laws, usually enacted by a legislative body.
- *The Constitution:* This is a body of fundamental principles according to which a nation or other organization intends to be governed.
- *Custom:* This is a widely accepted way of behaving or doing something that is specific to a society.

Environmental law has emerged as a distinct and unique branch of law under the influence of ecology. It has been developed in response to growing concern over issues impacting the environment worldwide. Such issues include climate change, flooding, waste management, drought, flooding, air pollution, water pollution, degradation of soil and vegetation, and exploitation of natural resources. Most violations of environmental laws are a civil offense, but there are also criminal penalties for serious offenders.

11.3 ENVIRONMENTAL LAWYERS

Environmentalists are often regarded as doomsayers, warning about the possibility of ecological catastrophe. Environmental law attracts lawyers who are interested in how we impact the geology and biodiversity of the planet. Environmental lawyers often shape environmental law in the legislative branch. They perform various functions, often helping to shape governmental and corporate policies and actions on a national and international level.

An environmental lawyer job description typically includes the following duties [5].

- Analyze data from findings, cases, trials, and other sources
- Advocate for environmental regulations and protections
- Recommend corrective action and fines for offenders

- Effectively communicate how an event or plan may negatively impact humans or wildlife
- Provide counsel to clients
- Determine if there is enough evidence to represent a business, individual, or government agency
- Assess damages and impact from an event
- Interview clients and other people of interest to compile evidence to develop a case
- Effectively and persuasively present evidence and other findings at trial
- Provide legal advice and assistance to investigators during criminal investigations
- Support and mentor junior environmental lawyers
- Provide legal and policy counsel to corporations, agencies, and other entities
- Act as a mediator between landowners and businesses or government agencies
- Draft environmental policies and business practices
- Organize and participate in educational campaigns and lobbying efforts
- Create clearly written communication to stakeholders and interested public parties
- Consult and advise agencies, researchers, scientists, and individuals
- Encourage public engagement by organizing environmental advocacy programs and forums

Figure 11.1 An environmental lawyer at work [6].

Figure 11.1 shows an environmental lawyer at work [6]. Environmental lawyers work for either public or private agents. They work as in-house counsel in a company or represent individuals in a formal court. They also work for administrative bodies, such as the Environmental Protection Agency, the Department of the Interior, the Department of Agriculture, and their state-level equivalents.

11.4 CHARACTERISTICS OF ENVIRONMENTAL LAW

Here we discuss the characteristics that set environmental law from other areas of law.

First, environmental law is paradoxical. Although environmental law functions like other kinds of law in establishing concepts, procedures, and rules to resolve conflicts, environmental law addresses the challenges of environmental issues. Second, environmental law must cope with the difficulties presented by time and space. Space can affect the ability of the individual seeking justice for an environmental harm [7]. Third, environmental law involves governance structures that are constructed to make choices about land-use and the social choices we make about the use. Fourth, much environmental law is administrative law because governments set environmental standards, to make land-use decisions, and also to enforce these regimes. Fifth, environmental law applies to environmental problems and relates to property rights in a unique way. The rights in land use inter-relate and property right-holders do not have absolute dominion over their private spaces [8].

Environmental law is regarded as a relatively new area of law. It is an interdisciplinary field covering various disciplines such as religion, philosophy, ethics, science, economics, national and international jurisprudence. It uses science (especially biology, chemistry and physics) to predict and regulate the consequences of human behavior on natural environment [9]. While science claims objective facts as reality, law seeks to reconcile disputes about facts. The law is often considered as over technical, and not sufficiently informed about science. The intersection of science and environmental law reflects the multidisciplinary nature of environmental law [10].

Some key functions of environmental law include [11]:

- Establish regulatory structures for environmental management, including regulatory agencies such as the EPA.
- Enable regulators to manage environmental impacts using plans, policies, standards, licenses, and incentives.
- Require those proposing environmentally significant activities to obtain approval from regulators.
- Enable members of the public to take part in environmental planning and environmental assessment.
- Require activities of environmental significance to be assessed before permission can be granted.
- Provide administrative, civil, and criminal penalties for breaches of the law.
- Allow the legality or merits of the certain decisions of regulators to be challenged by members of the public in appropriate courts.
- Allow for third parties to enforce breaches of the law through court action.

11.5 PRINCIPLES OF ENVIRONMENTAL LAW

Modern environmental law is shaped by a set of core principles, which include the following [12]:

- *The sustainability principle*: There is increasingly a need for laws concerning sustainability. Sustainable development necessitates balancing the concerns of both the environment and the economy. The environmental law serves as a tool to balance these potentially conflicting interests.
- *The precautionary principle:* This requires that it is better to control that activity now rather than wait for incontrovertible scientific evidence.
- *The prevention principle:* It is cheaper, easier, and less to prevent environmental harm or disaster than reacting to it when it happens.
- *The "polluter pays" principle:* The polluter-pays principle guides environmental regulations. Since the early 1970s, the "polluter pays" principle has been a dominant in environmental law. The main objective of many environmental regulations is to force polluters to bear the real costs of their pollution.

- *The integration principle:* Several jurisdictions and organizations have integrated environmental policies into their decision-making processes.
- *The public participation principle:* Public participation in environmental decision-making has been facilitated in the US bylaws. Policies on environmental protection often formally integrate the views of the public.
- *The proximity principle:* This requires that waste should be treated or eliminated if possible at the source. Waste should be recycled or burnt, but it should not be exported.

We should observe that the principles do not provide for a balancing of benefits and costs. This omission may cause waste of resources.

11.6 SCOPE OF ENVIRONMENTAL LAW

There are many areas under the umbrella of environmental law. Environmental laws have the following components [13]:

- *Air quality laws*: These govern the emission of air pollutants into the atmosphere and protect the air from pollution. These laws enforce air standards through monitoring that determines what constitutes safe levels of certain substances emitted by industrial processes, motor vehicles, etc. Figure 11.2 illustrates combating pollution with environmental law [14].

Figure 11.2 Combating pollution with environmental law [14].

- *Water quality laws:* These protect water from pollution. Three-fifths of the earth's surface is covered in water. Water quality laws govern the release of pollutants into water resources, including surface water, ground water, and drinking water supply, and the water table, rivers, seasons, and oceans. They may also determine who can use water. The US EPA (Environmental Protection Agency) is responsible for monitoring and enforcing standards to ensure our waterways are as clean and as healthy as possible.

- *Waste management laws:* These laws cover municipal waste, hazardous substances, and nuclear waste. They govern the transport, treatment, storage, and disposal of all manner of waste material produced as a byproduct of any industrial activity. Waste is a fact of life and is sometimes unavoidable. Waste management is concerned with the governance of many aspects of waste from their transport and storage.
- *Environmental cleanup laws:* These laws address pollution after it happens. They govern the removal of pollutant or contaminants from environmental media. They cover any gas emitted into the atmosphere from industrial activity. Regardless of whether such a pollutant leak is avoidable or unavoidable, there are necessary laws that determine what is required of the responsible party for the cleanup. The laws may include criminal punishment for polluters.
- *Chemical safety laws*: These laws govern the use of chemicals in human activities, particularly man-made chemicals in modern industrial applications. They cover materials introduced into an environment that have negative or harmful effects to the ecology or reduces the efficiency or safety of a resource.
- *Water resources laws:* These govern the ownership and use of water resources, including surface water and ground water. These laws cover the process of harnessing and using water in areas where drought is likely or managing it for minimal wastage. When water is handled poorly, it can lead to shortages.
- *Mineral resource laws:* These cover several basic topics, including the ownership of the mineral resource and who can work with them. For example, mining is regulated concerning the health and safety of miners.
- *Forest Conservation laws*: These are regulatory assurances that forest practices meets the requirements. Forestry is regulated by state and federal environmental laws. Lawmakers determine who can hunt and fish and how these activities are regulated. They also determine who can use natural resources and on what terms. Laws may regulate the use of natural resources such as timbers, forest protection, mineral harvesting, and animal and fish populations.

11.7 INTERNATIONAL ENVIRONMENTAL LAW

Environmental issues such as climate change, desertification, species extinction, chemicals and waste, ozone depletion, water scarcity, loss of biological diversity, and pollution of air, land, and seas have resulted in a growing international awareness that necessitates the need for nations to act collaboratively and find effective solutions to these problems. Environmental law is the basis for environmental sustainability and its global realization is a growing global concern. International law is the "law of nations." Global or International environmental law (IEL) is the branch of public international law which addresses States and international organizations with respect to the environment. The primary objective of IEL is to protect the environment and manage natural resources as specified in the general laws of each nation. Such law consists of the common set of legal principles developed by national, international, and transnational environmental regulatory systems [15].

International law is the law generated by nations attempting to regulate the international transactions. It includes rules drawn from treaties, international trade and commerce, international agreements, the laws of war and peace, the law of the sea, the laws of diplomatic practice, the law of treaties, the law of recognition and succession of states and governments, international human rights law, etc. [16].

The history of international environmental law goes back to the beginning of the twentieth century. The development of international environmental law as a distinct component of public international law began with the Stockholm Conference on the Environment in 1972. Today, international environmental law (IEL) is made up of modem environmental norms, which have been developing since the Stockholm Conference. Just like environmental law, IEL is interdisciplinary as shown in Figure 11.3 [17].

Figure 11.3 International environmental law is interdisciplinary [17].

The commonality of legal principles, rules, and regulatory tools and approaches resulting from these influences has created a global environmental law. International environmental law provides the common platform in which international environmental lawyers from different nations can collectively understand the international environmental issues, requirements, and solutions confronting nations.

The key features of international environmental law include [18]:

1. States have sovereignty over their natural resources and the responsibility not to cause environmental damage.
2. States must cooperate with each other and be good neighbors.
3. In the absence of scientific consensus that an action is harmful, the burden of proof that it is not harmful falls on the person taking the action (the precautionary principle).
4. The party responsible for producing pollution is responsible for paying for the damage done to the environment (the "polluter pays" principle)
5. Sustainable development – this concept can be found expressly and implicitly in many environmental treaties and can be considered as development that meets the needs of the present without compromising the ability of future generations to meet their own need.

11.8 SCOPE OF INTERNATIONAL LAW

The scope of international law is vast. It includes the environmental issues of population, biodiversity, global climate change, ozone depletion, management of toxic and hazardous substances, land/air/water pollution, dumping, conservation of marine living resources, desertification, nuclear damage, etc. The components of international environmental law include treaties, agreements, and soft law, and a few customary norms.

- *Environmental Treaties:* Treaties govern many aspects of international environmental law, e.g. ozone treaties. They are also known as conventions and protocols; they are the primary source of international environmental law. A treaty is a legal instrument which is intended to create legal rights and obligations between the parties. It goes through the process of negotiation, signature, and ratification. There are thousands of treaties (agreed to by governments and enforced) out there designed to save the world. They are governed by international law and binding once entered into force.
- *Environmental Agreements*: By nature, most environmental problems can only be addressed effectively through international cooperation. International environmental agreements are a

category of signed treaties with political and economic ramifications designed to regulate human impact on the environment in an effort to protect it. They can be bilateral agreements between two nations (sharing border) such as negotiating international environmental disputes between them in good faith. For example, a bilateral environmental agreement (BEA) between Finland and Russian Federation on cooperation in environmental protection. The agreement can also be multilateral environmental agreements (MEAs) between two or more nations. The multilateral agreements may be negotiated under the auspices of the UN. Common examples are Kyoto protocol on climate change and Montreal protocol on substances that deplete the ozone layer. There are hundreds of international environmental agreements most of which link only a limited number of countries. Some modern MEAs have almost universal membership.

- *Soft law*: This refers to non-binding documents which have the appearance of law, e.g. memorandum of understanding. Soft law has emerged in regions and worldwide since the end of the Second World War. Hard law comes mainly from custom or treaties, which take a long time to negotiate. A much more politically attractive approach is the "soft law." Soft law norms are used extensively in international trade law [19]. Soft law may evolve into hard law. While hard international law are generally enforceable by a national or international body, soft international law by itself is not enforceable.
- *Customary laws:* Customary laws play a secondary role in international environmental law. Indigenous peoples' customary laws are often used to criticize regulatory approaches to environmental management. The interactions between environmental law and the customary laws have attracted the attention of comparative lawyers. It has also been noted that non-governmental organizations (NGOs) actively support creative linkages between communities' customary law and international norms on sustainable development [20]. Customary international law can be developed even when the states do not fully comply with a particular norm.

Environmental responsibilities are shared among some specialized agencies such as Food and Agriculture Organization, World Health Organization, the International Maritime Organization, and the UN Development Programme (UNEP). UNEP addresses environmental issues at the global and regional levels. There are many international organizations which are involved in developing international environmental laws, regulations, and agreement. These include:

1. *United Nations* (UN): UN is involved in MEAs worldwide on a number of issues, including biological diversity, chemicals and waste, and climate and the atmosphere.
2. *World Trade Organization* (WTO): The WTO has been involved in MEA negotiations due to the agreements' trade implications.
3. *Non-governmental organizations* (NGOs): These are broadly recognized as providing important and useful contributions to the vast majority of environmental debates at the international level. The participation of such organizations in IEL reflects the ideals of participatory democracy, transparency, and socially inclusive globalization [21].

11.9 PRINCIPLES OF INTERNATIONAL ENVIRONMENTAL LAW

Principles play an important role in international environmental law. They generally operate in the evolutionary regulatory regime required to meet some challenges. They may have international legal significance. There is a proliferation of principles in the field of international environmental law. Here we discuss some key principles of international environmental law, similar to those in Section 11.5 [22]:

- *The Precautionary Principle:* This is an important legal principle to cope with the environmental risk of scientific uncertainty. It establishes legal mechanisms to protect the environment when scientific development leads to the uncertainty between the human activities and environmental damage. It generally means that if the adverse effect of anything is not determinable, it is better not to introduce it into the environment. The principle has become a policy-making framework, which is sensitive to environmental protection. It is being adopted in domestic laws for protecting human health [23]. It is regarded as a fundamental tool to achieve sustainable development.
- *The Principle of Protection:* This principle has a strong presence in international texts and practices. Environmental protection implies abstaining from harmful activities and adopting affirmative measures to prevent environmental degradation.
- *The Principle of Sustainable Development*: This principle reconciles three pillars: economic development, social equity, and environmental protection by adopting a developmental path.
- *The "Polluter Pays" Principle*: This supplies the means by which the cost of pollution prevention, control, and reduction measures are borne by the polluter. The overarching principle is recognized as an integral component of sustainable development.
- *The Principle of Participation:* A broad understanding of environmental justice involves participation in environmental controversies. The access rights are the core elements of the principle of participation.
- *Differential Treatment:* This is emerging legal principle. In a world where all are equal, but some are more equal than others, it is difficult to argue that all nations should have the same obligations to prevent environmental degradation [24].
- *The Forest Principles:* They are non-legally binding authoritative statement of principles for a global consensus on the conservation and sustainable development of all types of forests.

11.10 APPLICATIONS OF ENVIRONMENTAL LAWS

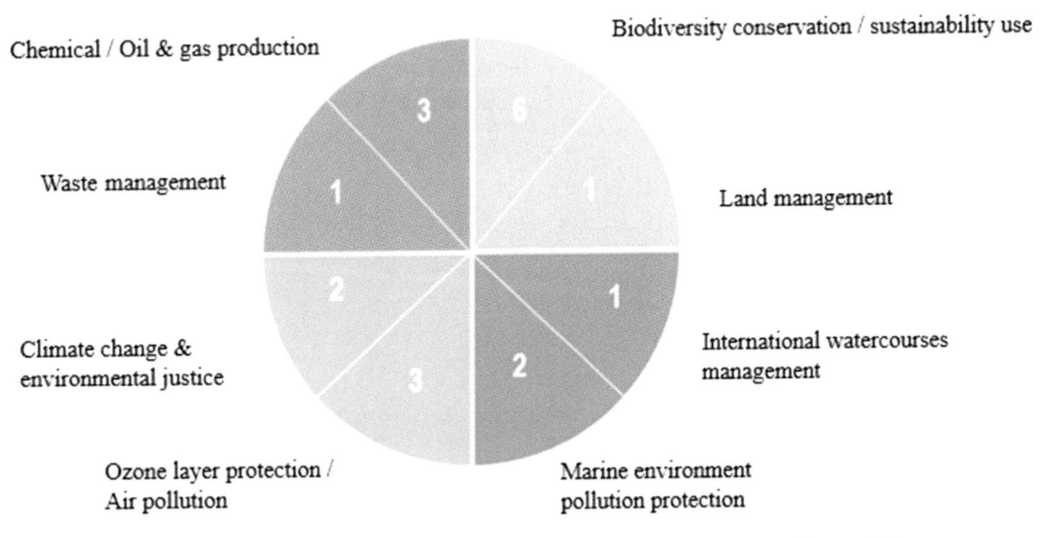

Figure 11.4 Some application areas of environmental law [25].

As shown in Figure 11.4, environmental law can be applied in several areas such as biodiversity conservation, land management, international watercourses, ozone layer protection, climate change, waste management, chemical management, manufacturing pollution, farming, and water pollution [25]. Some of these are covered briefly here.

- *Electronic Manufacturing Pollution:* Although electronic industries have played a vital role in driving the world economy, manufacturing of high-tech products has potential risks of releasing chemicals and generating tons of wastewater daily. Environmental laws can identify and regulate the pollutants generated from the electronic manufacturing [26].
- *Farms and Farming*: Farming is closely linked with human civilization. It has a dramatic impact on our environmental systems. Farmers apply fertilizers, nitrogen, phosphorous, and potassium to promote crop growth. Animal waste is major source of air pollution. Farms are virtually unregulated by the expansive body of environmental law developed in the US in the past 30 years. We should not ignore the pressing need for environmental regulation of farms [27].
- *Water Pollution:* The major law governing pollution of the nation's surface waters is the Clean Water Act, enacted in 1948 and revised in 1972. Its objective was to restore and maintain chemical, physical, and biological integrity of the nation's waters. The act utilizes water quality standards to protect water quality. It provided state and local governments with funds to address water pollution problems [28].
- *Waste Management:* How a waste is managed depends on whether it is a "solid waste" or a "hazardous waste." The management of solid waste is left primarily to individual states. EPA is mainly responsible for implementing the hazardous waste program. EPA established standards that include regulations on record keeping and reporting, waste accumulation time limits, and storage requirements [29].

International environmental law is applicable to numerous matters relating to air emissions, water, carbon, climate change, maritime pollution, oil and gas production, war, health, and a host of other things.

- *Human Health:* It has long been recognized that human health depends on nonpolluted, balanced natural ecosystems on local and global scales. Environment is often a source of ill-health to many people, especially in the developing nations. To move toward sustaining human health, international communities must include short- and long-term environmental health costs into economic globalization activities [30]. Many MEAs have health implications, e.g. Montreal protocol.
- *Climate change:* This is also called global warming. It is unquestionably the significant global environmental challenge. The primary cause of climate change is the burning of fossil fuels, such as oil and coal, which emits carbon dioxide. Impacts of climate change include the loss of sea ice, accelerated sea level rise and longer, and more intense heat waves. The umbrella for a global climate change response is provided by the United Nations Framework Convention on Climate Change. The agreements reached at Kyoto are truly revolutionary. Kyoto Protocol contained clear, precise, prescriptive and deadline-driven obligations backed up by a compliance system with enforcement power.
- *Water Management:* Water law principles have a long historical background. A challenge for the international community is the lack of access to safe drinking water, and inadequate sanitation. With increasing water scarcity, the management of shared resources between two or more sovereign states became an urgent issue in international law community. Water has also become a crucial focus of concern in international politics. Water projects have been taken over by giant transnational corporations [31].

- *Oil and Gas Production:* Offshore oil and gas exploration and exploitation have been increasing worldwide in recent years. There is no global international convention devoted to the governance of marine installations like drilling structures and oil production platforms. Commonly applied environmental law consists of law of the sea and maritime law sources [32].
- *Armed Conflict:* Recent international conflicts have raised fundamental questions about the impact of war on the human environment. The majority of international treaties do not apply during armed conflict. As for the treaties that are in principle applicable during armed conflict, belligerent and neutral States have the legal right to suspend those treaties, wholly or partially. There remains a need for more rules on environmental standards for military operations.
- *Air Pollution:* National laws are still the predominant controls on air pollution in mineral development. The trend in air pollution control is increasingly restrictive of mineral processes, viewed as the primary sources of chemical pollutants.
- *Marine Pollution:* This is a relatively long-standing concern. Oceans can be destroyed through pollution. To avoid global threats, some strict measures to control pollution are necessary. Protection against pollution is thus a fundamental aspect of ocean management. This will eventually help to achieve environmental objectives of almost all MEAs. International environmental law dealing the protection and preservation of the marine environment is well-developed. The International Seabed Authority is responsible for providing effective protection for the marine environment. Among the environmental problems dealt with at Stockholm, Canada were those concerning the increasing pollution of the world's waters. The International Maritime Organization is the major international organization dealing with the regulation of marine pollution [33].

Other applications of environmental law include animal protection (e.g. sea turtles, shark, migratory birds, etc.), waste management, forestry, mining, environmental justice, and the attainment of peace and security. Some of these issues do not fit nicely within political boundaries.

11.11 ENVIRONMENTAL LAW IN DEVELOPED NATIONS

By the turn of the twentieth century, a pattern of environmental cooperation had already emerged among developed nations, and numerous bilateral and multilateral treaties had already been enacted. National environmental laws and practice are increasingly influencing MEAs and international law practice. For example, domestic environmental enforcement can be applied to imports at ports and border crossings.

- *United States*: Most US states recognize a constitutional human right to a healthful environment. US has experienced environmental lawmaking without Congress, which plays a pre-eminent role as the nation's lawmaker. In the 1970s and 1980s, one typically looked to North America for leading precedents and trends in environmental law. Policymakers have recently paid great attention to the criminal sanction as a tool for reaching environmental objectives. There are many state environmental laws and regulations that have been designed to work together with the federal programs. The Environmental Protection Agency (EPA), the Occupational Safety and Health Administration (OSHA), and the Food and Drug Administration (FDA) have recently begun to formulate policies that move away from this historic dichotomy. In spite of the substantial environmental harms caused by farms, environmental law has given them a virtual license to continue to do so [34].

- *United Kingdom:* Environmental legislation in the UK including Scotland is intricately connected to European law. The United Kingdom's decision to leave the European Union in 2016 has made substantive impact on environmental law of the UK. UK will lose many restrictions on its freedom to make its own environmental law, but it will not be free from all constraints. The devolution settlements enacted in 1998 has created legislative and executive bodies in Scotland, Wales, and Northern Ireland. With Brexit, control over environmental law will return to the UK and to the devolved administrations [35].
- *European Union* (EU): The Union (with members: Belgium, Denmark, Germany, Greece, Spain, France, Ireland, Italy, Luxembourg, Austria, Portugal, Finland, and Sweden) takes an active part in the elaboration, ratification, and implementation of multilateral environmental agreements. The key objectives of the EU (2014 – 2020) are to protect, conserve, and enhance the Union's natural capital and safeguard the Union's citizens from environment-related pressures and risks to wellbeing. The Union has relied on the communality among its states to encourage resource management of a commons through social pressure and a spirit of cooperation. The European Environmental Agency, based in Copenhagen, Denmark, monitors the compliance of individual European nations with environmental EU directives. Lobbying is a necessary element in the European Community's (EC's) legislative process and is generally welcomed by the EC's law-making institutions [37].
- *Canada*: Canada has taken an initiative due to the country's variety of natural resources, climates, and populated areas, all of which can contribute to environmental stress. The Canadian draft of declaration included nine operative principles for the conduct of global environmental relations. Four of the principles had strong international legal implications [38]: (1) Every state has a duty to conduct its activities with due regard to their effects upon the environment of other states; (2) No state may use or permit the use of its territory in such a manner as to cause damage to areas beyond the limits of national jurisdiction; (3) No state may use areas beyond national jurisdiction in such a manner as to cause damage to such areas; (4) Every state has a duty to consult with other states prior to undertaking activities, which may damage the environment of such states. Ottawa invested significant political and diplomatic energy in UNCHE (United Nations Conference on the Human Environment), aiming to increase the nature and scope of international action to control the effects of pollution.
- *Australia:* The impact of international environmental law on Australian law has grown significantly. Australia is noted for its diverse environment law, which includes beaches, deserts, mountains, and climate change. There has been considerable debate over the power of the commonwealth to negotiate and adopt treaties and conventions [39].

11.12 ENVIRONMENTAL LAW IN DEVELOPING NATIONS

Over the years, there have been clear changes in developing nations, attitudes with respect to aspects of the environment falling within their territories. The market for water infrastructure and pollution prevention in the developing nations is massive. Developing nations either did not make water agreements at all, or did not effectively implement them.

- *India:* India has been actively involved in international negotiations, sometimes assuming a leadership role among developing nations. It has addressed the question of environmental human rights in the context of three green issues on water rights, forest conservation, and air pollution. In Indian environmental law, the "polluter pays" principle includes environmental costs as well as direct costs to people or property [40]. The Centre for Environmental Law (CEL) has compiled and analyzed existing international environmental agreements to which India is a party.

- *Bangladesh:* This country has been in vulnerable condition for environmental destruction over the years. Environmental laws of every nation evolve on the basis of different international initiatives and Bangladesh is no exception. The fight against the causes of environmental degradation started in Bangladesh after obtaining its independence from Britain in 1971. Bangladesh is attempting to develop its legal framework on the basis of international environmental law. The time is ripe for the government in Bangladesh to incorporate international environmental law in its legal system [41].
- *China*: Although the economy of China has developed rapidly, it is still under several severe pressures, including the elimination of poverty; the prevention and control of air, water, and soil pollution; and the prevention and control of ecological destruction. China has adopted regulations similar to those in other nations. Chinese environmental laws and regulations are abundant, but they suffer from a lack of enforcement. In 2003, China is the second-largest consumer and the third-largest producer of chemicals in the world. It adopted a set of regulations covering new chemical substances. It has embraced producer responsibility by requiring that all electronic information products must contain recyclability markings. Its environmental law in the 21st century continues to evolve in legislation, enforcement, higher education, and research. The Environmental Protection Law is the basic legislation for environmental protection in China [42].
- *Malaysia:* This is one of the very environmentally rich nations in the world. It is one of the fastest growing economies in the ASEAN region. At present, the high dense resourceful nation has been facing numerous environmental problems such as air pollution, water pollution, exploitation of natural resources, land degradation, deforestation, biodiversity degradation, depletion of environmental resources, wetland degradation, urban waste management, etc. The man made laws are not adequate to solve the environmental problems. The constitutional safeguards are not direct in protecting the environment. Some administrative bodies, such as the Department of Environment, the Ministry of Science, Technology and Environment are actively working on these issues [43]. The country has implemented several generations of elaborate schemes to promote agribusiness operations, which cause rapid conversion of undeveloped areas into plantation.
- *Taiwan:* Many high-tech parks have been established by the government Taiwan to promote national economic development. The high tech electronic companies have played a vital role in driving the world economy and providing jobs. However, the manufacturing of high-tech products release lots of chemicals and generate massive wastewater daily. High-tech industry-relevant pollution control regulations are still lacking. Environmental policies and regulations have failed to effectively monitor and manage pollution from chemicals used in the industry [44].

11.13 BENEFITS

Environmental law affects and benefits all of us— individual health, business activity, geographical sustainability. Environmental protection laws are designed with human health in mind. They are in place to preserve the status quo where they are beneficial and to tackle the harm being done for long-term sustainability. They are to protect the environment from harm and create rules for how people can use natural resources. They help ensure the environment and the economy are equally protected because each needs the other. Effective environmental laws can prevent decision-makers from rushing to approve projects that may harm our society, environment, and economy. They may regulate air/water pollution, flooding, the exploitation of natural resources, and waste management.

International environmental law ensures that activities within a jurisdiction do not cause any harm to the environment of other states. Its application has resulted in the development of environmental justice, which includes the issues of fairness, equity, standing, rights of disadvantaged populations in developing countries. Today, international law is not limited to agreements between the states. Matters of social concerns, such as health, education, economics, and human rights fall within the scope of international regulations [12]. Other benefits of IEL include reciprocity, cooperation, compliance, cost effectiveness, and participation.

11.14 CHALLENGES

The major challenge to the effectiveness of environmental laws is compliance due to the uncertain legitimacy of environmental law. For various reasons, environmental law has always been controversial. Debates over the necessity, fairness, scientific uncertainty, and cost of environmental regulation continue. Arguments against criminalizing environmental law are many. Environmental activist disagree with the benefits of environmental regulations.

Environmental assessment is being disreputed and criticized. Regulated or affected industry often argue against environmental regulation due to the cost involved. It is difficult to quantify the value of an environmental issue— such as a healthy ecosystem, clean air, or species diversity. Measures for environmental protection are costly. Another challenge is lack of trained lawyers to deal with regulating environmental quality. The integration of interdisciplinary perspectives in environmental law presents a great challenge for the educator.

The existing political structures constitute a major challenge to the creation of global environmental regulations. There are about 160 sovereign states in the world. Each state has economic sovereignty and the right to exploit and enjoy their natural resources without interference. Due to its sovereignty, no nation can be forced to participate, only encouraged to do so. Consequently, global structures to effectively regulate the negative environmental health consequences of globalization are not yet to be developed. There is the perception that the international environmental process is insufficiently democratic. The international environmental law has been criticized as toothless because enforcement power is generally weak. There is considerable debate among international legal scholars on the appropriate course of action necessary to avert global warming. There is always sources of conflict and uphill battle between environmentalist and anti-environmentalist organizations. The existing processes of making international environmental law are slow, cumbersome, expensive, ineffective, insufficiently democratic, and poorly coordinated. In spite of the numerous international environmental conventions, treaties, and agreements, which are accepted as international law; these laws are met with low levels of compliance and are ineffective because nations have been unwilling to cede their sovereignty. Other challenges include: coping with the proliferation of negotiated instruments; overcoming political opposition to environmental commitments; sorting out the relationship between environmental ethics, science and the rule of law; fleshing out the principles of sustainable development; and addressing practical problems of implementing international responsibilities [42].

11.15 CONCLUSION

Environmental law deals with some of today's most challenging environmental issues. It has come a long way in the last 30 years. Environmental policies must promote our highly valued national goal of a healthy environment while encouraging innovation and high-skilled, high-paying jobs and staying competitive in the global market. We should let environmental ethics influence our actions.

International environmental law may be regarded as a body of law concerned with protecting the environment through bilateral and multilateral international agreements.

It has generally failed to halt or reverse the rapid deterioration of the planet's ecosystems.

The development of international environmental law is a dynamic process. It requires continuously keeping abreast of not only current, but also future environmental trends and challenges. In order to address international environmental issues, it is imperative and expedient to initiate action at all levels—global, regional, national, local, and community. Although nations have been successful in negotiating international agreements, they are behind in making the agreements operate effectively [43].

Education that provides an informed basis for an understanding of environmental law is necessary to increase awareness of environmental issues. Some academic institutions have started introducing environmental law courses into the undergraduate curriculum.

One approach is to divide the courses generally into two categories: planning law and pollution controls [44]. More information on environmental law can be found in numerous books in [25,45-130] and several other books available in online marketing platforms. One may also consult the following related journals:

- *Asia Pacific Journal of Environmental Law*
- *Chinese Journal of Environmental Law*
- *Columbia Journal of Environmental Law*
- *Environmental Law*
- *Environmental Law Review*
- *Harvard Environmental Law Review*
- *Hastings Environmental Law Journal*
- *Journal of Environmental Law and Litigation*
- *Journal of Environmental Law and Practice*
- *Journal of Land Use and Environmental Law*
- *Journal of Property, Planning and Environmental Law*
- *Stanford Environmental Law Journal*
- *Tulane Environmental Law Journal*
- *Vermont Journal of Environmental Law*
- *Virginia Environmental Law Journal*
- *The American Journal of International Law*
- *International Environmental Agreements: Politics, Law and Economics*
- *Review of European Community & International Environmental Law* (RECIEL)
- *Yearbook of International Environmental Law*
- *Journal of International Trade Law and Policy*
- *Journal of Transnational Law & Contemporary Problems*
- *European Journal of International Law*
- *International Environmental Law Reports*
- *International Journal of Marine and Coastal Law*
- *Stanford Journal of International Law*
- *Yale Journal of International Law*
- *Syracuse Journal of International Law and Commerce*
- *International Journal of Environmental Studies*

REFERENCES

[1] C. G. Gonzalez and S. Atapattu, "International environmental law, environmental justice, and the Global South," *Transnational Law & Contemporary Problems,* vol. 26, no. 2, June 2107, pp. 229-242.

[2] M. N. O. Sadiku, O. D. Olaniyi, and S. M. Musa, "Environmental law: A primer," *International Journal of Trend in Research and Development,* vol. 6, no. 6, December 2019.

[3] "Environmental Law," Yale Law School.

[4] T. P. van Reenen, R. Knight, and S, Kaske, "Environmental & sustainability studies (ESS)," University of the Western Cape.

[5] "Environmental lawyer,"

https://unity.edu/careers/environmental-lawyer/

[6] "Environmental law,"

https://www.foulston.com/what-we-do/environmental-law

[7] J. C. Gellers, "Greening critical discourse analysis," *Critical Discourse Studies,* vol. 12, no. 4, 2015, pp. 482-493.

]8] E. Scotford and R. Walsh, "The symbiosis of property and English environmental law – Property rights in a public law context," *The Modern Law Review,* vol. 76, no. 6, 2013.

[9] K. Ruppel-Schlichting, "Introducing environmental law," in O. C. Ruppel and E. D. K. Yogo (eds.), *Environmental law and policy in Cameroon - Towards making Africa the Tree of Life.* Nomos Verlagsgesellschaft mbH, 2018.

[10] J. McEldowney and S. McEldowney, "Science and environmental law: Collaboration across the double helix," *Environmental Law Review,* vol. 13, 2011, pp. 169-198.

[11] "What are the functions of environmental law?"

https://www.edonsw.org.au/hys_what_are_the_functions_of_environmental_law

[12] "Principles of environmental law,"

https://www.britannica.com/topic/environmental-law/Principles-of-environmental-law

[13] "Environmental law," *Wikipedia,* the free encyclopedia,

https://en.wikipedia.org/wiki/Environmental_law

[14] Z. Liu, "The people vs. pollution: Empowering NGOs to combat pollution with environmental law," August 2018,

https://www.newsecuritybeat.org/2018/08/people-vs-pollution-empowering-ngos-combat-pollution-environmental-law/

[15] 529. M. N. O. Sadiku, O. D. Olaniyi, and S. M. Musa, "International environmental law: A tutorial," *International Journal of Trend in Research and Development,* vol. 6, no. 6, December 2019.

[16] Mark W Janis, "International law and treaties," in *International Encyclopedia of the Social & Behavioral Sciences,* 2nd edition, 2015, pp. 507-510.

[17] "International environmental law,"

https://www.slideshare.net/namari11/international-environmental-law-presentation

[18] S. Valentine and R. Smith, "International environmental law,"

http://www.a4id.org/wp-content/uploads/2016/03/International-Environmental-Law.pdf

[19] G. Palmer, "New ways to make international environmental law," *The American*

Journal of International Law, vol. 86, no. 2, April 1992, pp. 259-283.

[20] E. Morger, "Global environmental law and comparative legal methods," *RECIEL*, vol. 24, no. 3, 2015, pp. 254-263.

[21] A. Gillespie, "Facilitating and controlling civil society in international environmental law," *RECIEL*, vol. 15, no. 3, 2006, pp. 327-338.

[22] G. N. Gill, "The national green tribunal of India: A sustainable future through the principles of international environmental law," *Environmental Law Review,* vol. 16, 2014, pp. 183–202.

[23] A. H. Ansari and S. Wartini, "Application of precautionary principle in international trade law and international environmental law: A comparative assessment," *Journal of International Trade Law and Policy,* vol. 13, no. 1, 2014, pp. 19-43.

[24] L. Rajamani, *Differential Treatment in International Environmental Law.*

Oxford University Press, 2012.

[25] *International Environmental Law: Multilateral Environmental Agreements.* Hanoi, International Publishing House, May 2017.

[26] D. M. Bearden et al., "Environmental laws: Summaries of major statutes administered by the Environmental Protection Agency,"

https://digital.library.unt.edu/ark:/67531/metadc93953/

[27] W. Tu and Y. Lee, "Ineffective environmental laws in regulating electronic manufacturing pollution: Examining water pollution disputes in Taiwan,' *IEEE International Symposium on Sustainable Systems and Technology*, May 2009.

[28] J. B. Ruhl, "Farms, their environmental harms, and environmental law," *Ecology Law Quarterly*, vol. 27, no. 2, 2000, pp. 263-350.

[29] K. Knowlton, "Globalization and environmental health," in *Encyclopedia of Environmental Health*, 2nd edition, 2019, pp. 325-330.

[30] H. Elver, "International environmental law, water and the future," *Third World Quarterly*, vol. 27, no. 5, 2006, pp. 885-901.

[31] V. S. Radovich, "Oil and gas in the ocean- International environmental law and policy," *OCEANS 2016,* April 2016.

[32] S. A. J. Boelaert-Suominen, "International environmental law and Naval war: The effect pf marine safety and pollution conventions during international armed conflict,"

Doctoral Dissertation, University of London, April 1998.

[33] G. J. Matthews, "International law and policy on marine environmental protection and management: Trends and prospects," *Marine Pollution Bulletin,* vol. 25, no. 1-4, 1992, pp. 70-73.

[34] J. B. Ruhl, "Farms, their environmental harms, and environmental law," *Ecology Law Quarterly,* vol. 27, no. 2, 2000, pp. 263-350.

[35] C. T. Reid, "Brexit and the future of UK environmental law," *Journal of*

Energy & Natural Resources Law, vol. 34, no. 4, 2016, pp. 407-415.

[36] D. Gillies, "Lobbying and European community environmental law,"

European Environment, vol. 8, 1998, pp. 175–183.

[37] M. W. Manulak, "Multilateral solutions to bilateral problems: The 1972 Stockholm conference and Canadian foreign environmental policy," *International Journal,* 2015, vol. 70, no. 1, 2015, pp. 4–22.

[38] D. R. Rothwell and B. Boer, "The influence of international environmental law on Australian courts," *Review of European Community & International Environmental Law,* vol. 7, no.1, April 1998, pp. 31-39.

[39] M. R. Anderson, "International environmental law in Indian courts," *Review of European Community & International Environmental Law,* vol. 7, no. 1, April 1998, pp. 21-30.

[40] H, S. Hasan and S. A. Rahaman, "Principles of international environmental laws: Application in national laws of Bangladesh," *Contemporary Issues in International Law*, 2018, pp 101-109.

[41] W. Canfa, "Chinese environmental law enforcement: Current deficiencies and suggested reforms," *Vermont Journal ff Environmental Law*, vol. 8, 2007,, pp. 159-193.

[42] N. Mohammad et al., "Potentialities and constraints of the environmental

law and policy in Malaysia to protect the environment: An empirical study for sustainable development," *Proceedings of IEEE International Summer Conference of Asi2a Pacific Business Innovation and Technology Management*, 2011.

[43] W. Tu and Y. Lee, "Ineffective environmental laws in regulating electronic manufacturing pollution: Examining water pollution disputes in Taiwan," *Proceedings of*

IEEE International Symposium on Sustainable Systems and Technology, May 2009.

[44] R. Hammer, "Integrating interdisciplinary perspectives into traditional environmental law courses," *Journal of Geography in Higher Education,* vol. 23, no. 3, 1999, pp. 367-380.

[45] D. E Fisher, *Environmental Law in Australia: An Introduction.* University of Queensland Press, 1980.

[46] L. Borzsák, *The Impact of Environmental Concerns on The Public Enforcement Mechanism Under EU Law; Environmental Protection In The 25th Hour.* Wolters Kluwer Law & Business, 2011.

[47] P. S. Menell (ed.), *Environmental Law.* Ashgate Publishing Co., 2002.

[48] L. A. Malone, *Environmental Law.* Aspen Publishers, 2003.

[49] U. Beyerlin and T. Marauhn. *International Environmental Law.* Hart Publishing, 2011.

[50] L. Krämer. (ed.), *European Environmental Law.* Ashgate Publishing Co., 2003.

[51] Y. Shigeta, *International Judicial Control of Environmental Protection; Standard Setting, Compliance Control, and The Development of International Environmental Law. by the International Judiciary.* Kluwer Law International, 2010.

[52] Z. J. B. Plater et al., *Environmental Law and Policy: Nature, Law, and Society.* Aspen Publishers, 4th edition, 2010.

[53] R. S. Abate, *Directory of Environmental Law: Education Opportunities at American Law Schools.* Carolina Academic Press, 2nd edition, 2008.

[54] D. Shelton, *International Environmental Law.* Brill-Nijhoff; 3rd edition, 2004.

[55] P. W. Birnie and A. E. Boyle, *International Law and the Environment.* Oxford, UK: Oxford University Press, 1994.

[56] C. Bell et al., *Environmental Law Handbook.* Bernan Press, 23rd edition,2016.

[57] P. Sands and J. Peel, *Principles of International Environmental Law.* Cambridge, UK: Cambridge University Press, 2012.

[58] R. J. Lazarus, *The Making of Environmental Law.* Chicago: The University of Chicago Press, 2004.

[59] M. Lee, *EU Environmental Law, Governance And Decision Making.* Hart, 2014.

[60] J. H. Jans, *European Environmental Law.* Europa Law Publishing, 2nd edition, 2000.

[61] N. de Sadeleer, *EU Environmental Law and The Internal Market.* Oxford University Press, 2014.

[62] P. Salmon and D. Grinlinton (eds.), *Environmental Law in New Zealand.* Thomson Reuters, 2015.

[63] S. C. Shastri, *Environmental Law in India*. Eastern Book Company, 2nd edition, 2005.

[64] P. Cullet, *Differential Treatment in International Environmental Law*. Ashgate Publishing Co., 2003.

[65] A. Gillespie, *International Environmental Law, Policy, and Ethics*. Oxford University Press, 2nd edition, 2014.

[66] B. J. Richardson and S. Wood (eds.), *Environmental Law for Sustainability: A Reader*. Hart Publishing, 2006.

[67] E. Fisher, *Environmental Law – A Very Short Introduction*. Oxford University Press, 2018.

[68] R. V. Percival et al. (eds), *Global Environmental Law at a Crossroads*. Cheltenham: Edward Elgar Publishing, 2014.

[69] R. Duxbury and S. Morton (eds.), *Blackstone's Statutes on Environmental Law*.

Oxford University Press, 5th edition, 2004.

[70] D. Leary and B. Pisupati (eds.), *The Future of International Environmental Law*. Tokyo, Japan: United Nations University, 2010

[71] M. Lee, *EU Environmental Law, Governance and Decision Making*. Hart, 2014.

[72] S. Coyle and K. Morrow, *The Philosophical Foundations of Environmental Law. Property, Rights and Nature*. Hart Publishing 2004.

[73] E. Scotford, *Environmental Principles and the Evolution of Environmental Law*. Oxford, UK: Hart Publishing, 2017.

[74] J. H. Jans, *European Environmental Law*. Europa Law Publishing, 2nd edition, 2000.

[75] F. McManus, *Environmental Law in Scotland: An Introduction and Guide*. Edinburgh University Press, 2016.

[76] T. F. M. Etty et al., (eds.), *The Yearbook of European Environmental Law*. Oxford University Press, 2005.

[77] M. Faure and G. Heine, *Criminal Enforcement of Environmental Law in the European Union*. Kluwer Law, 2005.

[78] R. L. Revesz, *Foundations of Environmental Law and Policy: Interdisciplinary Readers in Law*. Oxford University Press, 1997.

[79] P. G. G. Davies, *European Union Environmental Law: An Introduction to Key Selected Issues*. Ashgate, 2004.

[80] J. F. McEldowney and S. McEldowney, *Environmental Law and Regulation*. Blackstone Press, 2001.

[81] S. Atapattu, *Emerging Principles of International Environmental Law*. Transnational Publishers, 2006.

[82] E. Louka, *International Environmental Law: Fairness, Effectiveness and World Order*. Cambridge University Press,2006.

[83] D. R. Boyd, *Unnatural Law: Rethinking Canadian Environmental Law and Policy*. Vancouver, Canada: UBC Press, 2003.

[84] G. Winter (ed.), *European Environmental Law: A Comparative Perspective*. Dartmouth, 1996.

[85] M. Fitzmaurice, *Contemporary Issues in International Environmental Law*. Cheltenham, Edward Elgar, 2009.

[86] J. Ebbeson and P. Okowa (eds.), *Environmental Law and Justice in Context.* Cambridge, UK: Cambridge University Press, 2009.

[87] A. Boyle and D. Freestone, *International Law and Sustainable Development: Past Achievement and Future Challenge.* New York: Oxford University Press, 1999.

[88] R. E. Meiners and A. P. Morriss (eds.), *The Common Law and the Environment: Rethinking the Statutory Basis for Modern Environmental Law.* Lanham, MD: Rowman & Littlefield, 2000.

[89] J. Salzman and B. H. Thompson, *Environmental Law and Policy.* Foundation Press, 2003.

[90] D. R. Boyd, *Unnatural Law: Rethinking Canadian Environmental Law and Policy.* Vancouver, Canada: The University of British Columbia Press, 2003.

[91] N. de Sadeleer, *EU Environmental Law and the Internal Market.* Oxford University Press, 2014.

[92] P. Salmon and D. Grinlinton (eds.), *Environmental Law in New Zealand.* Thomson Reuters, 2015.

[92] B. Boer, R. Ramsay, and D. R. Rothwell, *International Environmental Law in the Asia Pacific.* Kluwer Law International, 2006.

[93] U. Beyerlin and T. Marauhn. *International Environmental Law.* Hart Publishing, 2011.

[94] L. Krämer. (ed.), *European Environmental Law.* Ashgate Publishing Co., 2003.

[95] Y. Shigeta, *International Judicial Control of Environmental Protection; Standard Setting, Compliance Control, and The Development of International Environmental Law by the International Judiciary.* Kluwer Law International, 2010.

[96] D. E Fisher, *Environmental Law in Australia: An Introduction.* University of Queensland Press, 1980.

[97] D. Shelton, *International Environmental Law.* Brill-Nijhoff; 3rd edition, 2004.

[98] P. W. Birnie and A. E. Boyle, *International Law and the Environment.* Oxford, UK: Oxford University Press, 1994.

[99] J. H. Jans and H. H. B. Vedder, *European Environmental Law: After Lisbon.* Europa Law Publishing, 4th edition, 2012.

[100] S. C. Shastri, *Environmental Law in India.* Eastern Book Company, 2nd edition, 2005.

[101] F. McManus, *Environmental Law in Scotland: An Introduction and Guide.* Edinburgh University Press, 2016.

[102] T. F. M. Etty et al., (eds.), *The Yearbook of European Environmental Law.* Oxford University Press, 2005.

[103] M. Faure and G. Heine, *Criminal Enforcement of Environmental Law in the European Union.* Kluwer Law, 2005.

[104] P. G. G. Davies, *European Union Environmental Law: An Introduction to Key Selected Issues.* Ashgate, 2004.

[105] S. Atapattu, *Emerging Principles of International Environmental Law.* Transnational Publishers, 2006.

[106] T. Koivurova, *Introduction to International Environmental Law.* London, UK: Routledge, 2014.

[107] G. Winter (ed.), *European Environmental Law: A Comparative Perspective.* Dartmouth, 1996.

[108] M. Fitzmaurice, *Contemporary Issues in International Environmental Law.* Cheltenham, Edward Elgar, 2009.

[109] A. Boyle and D. Freestone, *International Law and Sustainable Development: Past Achievement and Future Challenge.* New York: Oxford University Press, 1999.

[110] R.Wolfrum and N. Matz, *Conflicts in International Environmental Law.* Springer Science & Business Media, 2003.

[111] D. Bodansky, J. Brunnée, and E. Hey, *The Oxford Handbook of International Environmental Law.* Oxford, 2008.

[112] P. Sands and J. Peel, *Principles of International Environmental Law*

Cambridge University Press, 2012.

[113] D. Bodansky, *The Art and Craft of International Law.* Cambridge, MA: Harvard University Press, 2010.

[114] E. Louka, *International Environmental Law: Fairness, Effectiveness, and World Order.* Cambridge University Press, 2006.

[115] A. Kiss and D. Shelton, *International Environmental Law.* Transnational Publishers, 1991.

[116] U. Beyerlin and T. Marauhn, *International Environmental Law.* Oxford, UK: Hart Publishing, 2011.

[117] A. Kiss and D. Shelton, *Guide to International Environmental Law.* Martinus Nijhoff Publishers, 2007.

[118] P. M. Dupuy and J. E. Vinuales, *International Environment Law.* Cambridge University Press, 2nd edition, 2018.

[119] S. Atapattu, *Emerging Principles of International Environmental Law.* Brill, 2007.

[120] D. Shelton, A. C. Kiss, *Judicial Handbook on Environmental Law.* UNEP/Earthprint, 2005.

[121] E. Morgera, *Corporate Accountability in International Environmental Law.* Oxford University Press, 2009.

[122] R. V. Percival, J. Lin, and W. Piermattei, *Global Environmental Law at a Crossroads.* Edward Elgar Publishing, 2014.

[123] L. J. Kotzé, *Global Environmental Governance; Law and Regulation for the 21st Century.* Edward Elgar Publishing, 2012.

[124] P. L. Fitzgerald, *International Issues in Animal Law; the Impact Of International Environmental and Economic Law Upon Animal Interests and Advocacy.* Carolina Academic Press, 2012.

[125] B. H. Desai, *Institutionalizing International Environmental Law.* Transnational Publishers, 2004.

[126] P. Cullet, *Differential Treatment in International Environmental Law.* Ashgate Publishing Co., 2003.

[127] A. Gillespie, *International Environmental Law, Policy, and Ethics.* Oxford University Press, 2nd edition, 2014.

[128] P. Sands and J. Peel, *Principles of International Environmental Law.* Cambridge, UK: Cambridge University Press, 2012.

[129] R. V. Percival et al. (eds), *Global Environmental Law at Crossroads.* Cheltenham: Edward Elgar Publishing, 2014.

[130] D. Leary and B. Pisupati (eds.), *The Future of International Environmental Law.* Tokyo, Japan: United Nations University, 2010

CHAPTER 12

ENVIRONMENTAL JUSTICE

"Environmental justice is the movement to ensure that no community suffers disproportionate environmental burdens or goes without enjoying fair environmental benefits." – Van Jones

12.1 INTRODUCTION

We all have a stake in our environment. We must ensure that our air is cleaner, our water is purer, and our land is protected. Environmental discrimination has historically been evident in the process of selecting where to dump waste and hazardous materials. It is well believed that environmental injustice is the outcome of an institutional oppression and isolation. Such environmental injustice has implications for human health and wellbeing, affecting the ability of present and future generations to be productive [1].

Environmental justice (EJ) is a critical part of social justice because environmental inequalities, like other forms of social inequalities, worsen health, hamper economic performance, diminish social cohesion, and promotes racism. The exact meaning of "justice" in "environmental justice" has remained contested because justice is a familiar but complex concept. There are apparent clashes between resource utilization, transport, and waste disposal. Increasing evidence shows that minority-based environmental injustice exists. For example, economically disadvantaged kids live in crowded homes and are exposed to environmental toxins.

From international viewpoint, environmental justice must be understood in terms of socially-derived inequalities in benefits and burdens of production systems. EJ often occurs between the global South and the global North. For example, communities in rural Cuba may be affected by a Canadian mining company. Another example is the case of an industrialized nation dumping their waste to a developing nation. Environmentalists or conservationists, activists, communities, humanitarian actors, and non-governmental organizations (NGOs) are demanding for much more than equity or just distribution.

This chapter provides an introduction on environmental justice. It begins by explaining the concept of environmental justice and providing some causes of environmental injustice. It discusses environmental justice principles and Environmental Justice Movement. It addresses the globalization of the environmental justice. It provides some applications of environmental justice. It covers how environmental justice is practiced in developed and developing nations. It highlights the benefits and challenges of environmental justice. The final section concludes with some comments.

12.2 CONCEPT OF ENVIRONMENTAL JUSTICE

Environmental justice or rather injustice is the fair treatment and inclusive participation of all people in environmental decision-making regardless of race, color, or socioeconomic status. The merger of ethics and environmental law forms the foundation of environmental justice, which mainly addresses the issue of environmental discrimination or racism. Scholars and activists for environmental justice argue that all people deserve to live in a clean and safe environment. Governments or civil society leaders should provide leadership to champion environmental justice by addressing grassroots concerns on environmental inequalities [2].

Environmental justice is a concept that emerged in the United States in the early 1980s, in the context of the struggle for racial equality. It encompasses the distribution of environmental goods and harms. It may also be regarded as a principle of American democracy that combines civil rights with environmental protection. Evidence indicates certain racial and ethnic groups bear a disproportionate burden of environmental hazards. EJ that must be considered in relation to the social, economic, and political factors that influence the experiences of affected communities. EJ tries to shed environmental racism. Several environmental justice issues revolve around issues of environmental risk.

Environmental justice is a multifaceted concept. It is also a "contested" concept which is susceptible to multiple interpretations. It is both a social movement and a framework through which to evaluate domestic and international laws, policies, and practices that have a disparate impact on vulnerable communities. Environmental justice scholars and activists have underscored four distinct features of environmental justice [3]:

(1) *Distributive justice:* This calls for the fair distribution of the benefits and burdens of economic activity, as well as equitable access to environmental goods and services. Distributive justice concerns the principles and processes for sharing benefits and harms. It is closely related with issues of racism, sexism, classism, and poverty.

(2) *Procedural justice:* This involves procedural equity and inclusiveness, including the right of all communities to participate in governmental decisions related to the environment. Individuals or communities are unable to participate in environmental decision-making.

(3) *Corrective justice*: This is the even-handed enforcement of environmental laws and the compensation of those whose rights are violated.

(4) *Social justice:* EJ is deeply connected with social and economic justice and cannot be achieved without addressing related social problems, such as poverty, racism, and human rights.

Without addressing these features (social, distributive, corrective, and procedural aspects), it would be difficult to rectify prevailing environmental injustices.

EJ is the principle that environmental costs and amenities ought to be equitably distributed within society regardless of race, color, or income. It is a concept whereby individuals, environment, and health are integrated to advance social policy and improve health for vulnerable populations. The rise of green movement in the 1960s helped to develop environmental consciousness and elevated the scholarly debate about environmental justice.

Environment justice is a global issue that concerns all nations. Access to the rights and burdens of the environment is unfairly distributed according to age, race, socioeconomic status, ethnicity, and geographical location within nations. At a global scale, environmental justice can apply to situations such as industrialized countries exporting their wastes to developing nations.

12.3 CAUSES OF ENVIRONMENTAL INJUSTICE

Environmental injustice is essentially the unequal distribution of benefits and burdens. Environmental burdens and benefits have a profound impact on people's capabilities. Environmental injustice is a matter of disproportionate impact. It exists when an individual or group of individuals is denied access to environmental investments, denied access to information, or disallowed from participating in decision-making.

The main causes of environmental injustice include the following [4,5]:

- *Institutionalized racism:* Racial discrimination in various forms contributes to environmental injustice, the unequal distribution of benefits and burdens. Environmental racism may be regarded as the placement of health-threatening structures in areas where the poor and ethnic minorities live. It refers to the institutionalized practices of government or corporate decision makers who deliberately exclude communities of low socio-economic status from making environmental decisions affecting them. Environmental racism may be regarded as the placement of health-threatening structures such as landfills and factories near or in areas where the poor, ethnic minorities live. It violates the prohibition against unequal protection of toxic and hazardous waste exposure. It was exploited by some to benefit another. This racism has become acculturated and engrained in contemporary social institutions. For example, pollution-producing facilities are often sited in poor communities of color. Environmental justice is an important part of the struggle to maintain a healthful environment, especially for those who have traditionally lived, worked, and played closest to the sources of pollution. Figure 12.1 shows racism as a cause of environmental injustice [6].

Figure 12.1 Racism as a cause of environmental injustice [6].

- *Commoditization of resources:* This is protecting environmental goods and resources such as water, energy, land, air, and green spaces for the benefit of those in power over those who lack power. There are resource extraction conflicts, transport conflicts, and also waste disposal conflicts. Disadvantaged residents are less likely to have the necessary political resources and power to fight against environmental injustice. The environmental justice emphasizes the disproportional proximity of low-income and minority communities to activities and contaminated land. It addresses issues such as food security, affordable transit systems, healthy housing, and responses to climate change. A just society is one in which everyone gets a fair share of the available resources.

- *Unresponsive government authorities:* Government authorities have often been unresponsive to community needs regarding environmental inequities. This has led to unresponsive and unaccountable governmental policies and regulations which exist at all levels of government. Environmental justice activists have pushed their agenda within government so that they can make environmental justice part of their environmental policy.
- *Powerlessness of the victim:* There is lack of power by the victims of environmental injustice. This is manifested by the ability of the victim to have few financial resources invest in the struggle for environmental justice. Environmental justice, as a principle of American democracy, demands that those who have historically been excluded from environmental decision making, traditionally minority and tribal communities, have the same access to environmental decision makers, decision-making processes as any other individuals.
- *Capitalism*: This is an economic, political, and social system in which property, business, and industry are privately owned, directed toward making the greatest possible profits. In the United States, capitalism has given birth to modern-day empires and billionaires. Several companies such as Chevron and Dominion Energy continue to profit from an environmentally harmful industry and their investment in climate change are not rare.

Typical practices of environmental injustice are [7]:

- The placing of hazardous and other noxious facilities
- Lead poisoning among children
- Asthma and other respiratory illnesses
- Unsafe, indecent, and exploitative workplace conditions
- Cancer, birth defects, and developmental illnesses
- Pesticide poisoning of farm workers
- Contaminated sites and properties
- Transportation thoroughfares
- Congested and decaying housing conditions
- Lack of protection of spiritual grounds and indigenous habitats
- Pollution and lack of sound economic development
- Lack of access to quality healthcare
- Unequal enforcement of environmental laws
- Lack of people of color in the environmental professions
- Inadequate community participation in the decision-making process

12.4 ENVIRONMENTAL JUSTICE PRINCIPLES

Environmental justice embraces the major principle that all people are entitled to equal protection of our environmental laws. The following seventeen principles have served as a defining document for the growing grassroots movement for environmental justice [8,9].

1. Environmental justice affirms the sacredness of Mother Earth, ecological unity and the interdependence of all species, and the right to be free from ecological destruction.
2. Environmental justice demands that public policy be based on mutual respect and justice for all peoples, free from any form of discrimination or bias.
3. Environmental justice mandates the right to ethical, balanced, and responsible uses of land and renewable resources in the interest of a sustainable planet for humans and other living things.
4. Environmental justice calls for universal protection from nuclear testing, extraction, production and disposal of toxic/hazardous wastes and poisons and nuclear testing that threaten the fundamental right to clean air, land, water, and food.

5. Environmental justice affirms the fundamental right to political, economic, cultural, and environmental self-determination of all peoples.
6. Environmental justice demands the cessation of the production of all toxins, hazardous wastes, and radioactive materials, and that all past and current producers be held strictly accountable to the people for detoxification and the containment at the point of production.
7. Environmental justice demands the right to participate as equal partners at every level of decision-making including needs assessment, planning, implementation, enforcement, and evaluation.
8. Environmental justice affirms the right of all workers to a safe and healthy work environment, without being forced to choose between an unsafe livelihood and unemployment. It also affirms the right of those who work at home to be free from environmental hazards.
9. Environmental justice protects the right of victims of environmental injustice to receive full compensation and reparations for damages as well as quality healthcare.
10. Environmental justice considers governmental acts of environmental injustice a violation of international law, the Universal Declaration On Human Rights, and the United Nations Convention on Genocide.
11. Environmental justice must recognize a special legal and natural relationship of Native Peoples to the US government through treaties, agreements, compacts, and covenants affirming sovereignty and self-determination.
12. Environmental justice affirms the need for urban and rural ecological policies to clean up and rebuild our cities and rural areas in balance with nature, honoring the cultural integrity of all our communities, and providing fair access for all to the full range of resources.
13. Environmental justice calls for the strict enforcement of principles of informed consent, and a halt to the testing of experimental reproductive and medical procedures and vaccinations on people of color.
14. Environmental justice opposes the destructive operations of multi-national corporations.
15. Environmental justice opposes military occupation, repression, and exploitation of lands, peoples and cultures, and other life forms.
16. Environmental justice calls for the education of present and future generations which emphasizes social and environmental issues, based on our experience and an appreciation of our diverse cultural perspectives.
17. Environmental justice requires that we, as individuals, make personal and consumer choices to consume as little of Mother Earth's resources and to produce as little waste as possible; and make the conscious decision to challenge and reprioritize our lifestyles to insure the health of the natural world for present and future generations.

12.5 ENVIRONMENTAL JUSTICE MOVEMENT

The social values of several local environmental justice activists have translated to the Environmental Justice Movement (EJM). The EJ movement grew in US in 1970s as a grassroots movement from concerns that hazards, such as toxic waste disposal facilities, were predominantly located in low income and non-white communities. Environmental Justice movement began in the United States and spread to other countries and became a global movement known as Transnational Networks for Environmental Justice [10].

Environmental justice advocates keenly contend that there is a relationship between the trilogy of environmental racism, environmental discrimination, and environmental policymaking. The resultant impacts of pollution on human health and well-being are generally spread nonuniformly among the populace. Robert Bullard, African-American sociologist, is widely regarded the "father of the environmental justice movement" [11].

The movement aims at building a national and international movement of all peoples of color to fight the destruction and taking of our lands and communities. The goals of the movements include: ending institutional discrimination, eliminating environmental inequality, and pursuing social justice. Other issues tackled by EJ in the United States include race, racial justice, inequality, environmental degradation, and social liberation.

The movement has achieved a measure of success, particularly in getting government to respond to their concerns and ensuring federal intervention. A key determinant of success is the ability of the movement to continuously push for environmental laws and policies that benefit their community [12]. A group of EJ activists is shown in Figure 12.2 [13].

Figure 12.2 A group of EJ activists [13].

EJM plays a crucial role in redefining and promoting sustainability. Organizing internationally towards a common goal is necessary to build a global environmental justice movement. Social mobilizations are over resource extraction, land grabbing, water quality, pollution from oil extraction, mining, environmental degradation, or waste disposal.

The Environmental Justice Atlas (EJAtlas) is an international collaboration that tracks land and energy conflicts around the world. The EJAtlas was conceived as a means of integrating activist knowledge into a global platform that will be useful to activism, advocacy, and public education. The EJAtlas makes visible many environmental injustices and provides case studies, database, and evidence to support the environmental justice movement or environmentalists. Among the conflicts covered by the EJAtlas by April 2015 were those about mining, industrial extraction of fossil fuels, land conflicts, disposal of waste materials, and water management conflicts [14]. Another international organization is the Environmental Justice Organizations, Liabilities and Trade (EJOLT), an FP7 project supported by the European Commission.

12.6 GLOBAL ENVIRONMENTAL JUSTICE

There are an increasing number of environmental conflicts around the world. These include land grabbing, water justice, climate justice, food sovereignty, unregulated oil mining, trash economy, nuclear nightmares, resource extraction, transport, and waste disposal. The conflicts and resistance are essentially a response to the growth and changes in the social metabolism of industrial economies, i.e., the flows of energy and materials in the economy. Communities around the world are organizing and opposing the imposition of "development" projects and addressing the conflicts, especially political and economic forces that produce injustices. They bring value system contests into the open. These efforts from the grassroots often lead to myriad environmental justice organizations (EJOs) around the world. There have been many successful examples of stopping many "developing" projects or developing alternatives.

The increasing awareness of our interconnectedness has developed through the global organization of science and knowledge about environment. The globalization of the environmental justice (EJ) has extended the EJ approach in the United States beyond US and apply it to diverse movements struggling against similar problems around the globe. Environmental justice has increasingly served as a crucial rallying ground for social activism and political resistance beyond the US. The concept of global environmental justice (GEJ) is emerging as an important concept in the field of environmental economics and development. GEJ is a global concern addressing global environmental practices and policies that support inequitable distribution of environmental benefits and burdens [15]. Figure 12.3 shows the global environmental justice atlas [16].

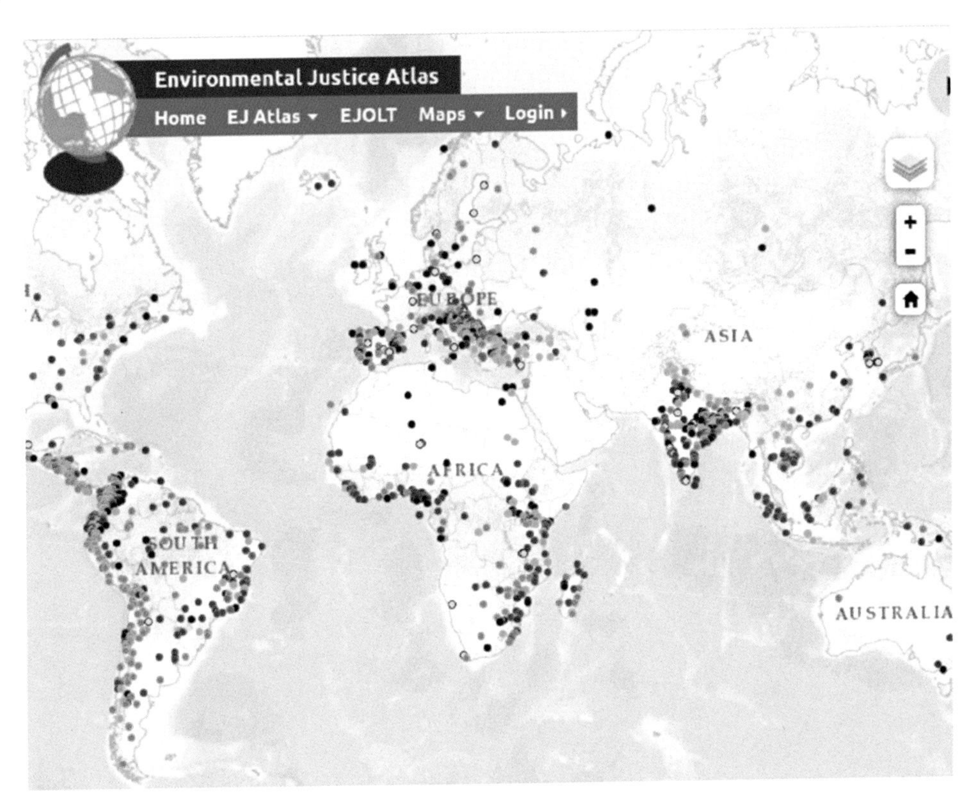

Figure 12.3 Global environmental justice atlas [16].

Since the language of justice is common, it is important to examine environmental struggles in justice terms. Whether we consider "climate justice," "food justice," or "water justice" the language of justice is omnipresent in environmental politics. Opposition to injustices lies at the heart of many struggles around the world. The so-called "rights-based approaches" have gained a prominent place in international struggles [17].

The global environmental justice must address the underlying drivers of environmental degradation [18]:

- *Population:* Analyses suggest that an escalation of proven noncoercive approaches could lead to a leveling off of global population at eight and a half billion people in this century.
- *Poverty and underdevelopment:* Poverty is an important contributor to environmental deterioration. Eliminating large-scale poverty is now considered a possibility.
- *Technology:* It is necessary to transform the technologies that today dominate manufacturing, energy, transportation, and agriculture. Transformation of the energy sector must rank as the highest priority.
- *Market signals*: Needed changes in technology and consumption patterns will not happen unless there is a parallel revolution in pricing. The corrective most needed now is environmentally honest prices. Doing the right thing environmentally should be cheaper, not more expensive!

12.7 APPLICATIONS OF ENVIRONMENTAL JUSTICE

Practicing environmental justice entails ensuring that all citizens receive from the government the same degree of protection from environmental hazards. The practice of EJ will be illustrated by the US government, US Department of Transportation, environmental health, and climate justice.

- *Government*: Environmental justice is regarded as a policy principle formulated by the government at local, state or federal level. The responsibility for drawing up these policies lies with the federal states, which can transfer responsibility to local agencies. Environmental justice initiatives now operate across multiple programs within the states in the US and the rest of the world. Achieving climate goals in the US disproportionately affects members of the population who are the poorest, most marginalized, and non-white. Fairness, justice, and equity must be the main aspects of our climate agenda. There is a relationship between the US prison system and environmental justice concerns. The government at all levels should have an inclusive agenda in ensuring social and environmental justice. As shown in Figure 12.4, the US President Clinton signed the executive order on EJ in 1994 [19]. The Environmental Protection Agency (EPA) in 2009 initiated activities to ensure the assessment and consideration of environmental justice issues in its regulatory decisions. Regulatory interventions by EPA was to fulfill its mission to protect the environment and health. Municipal government should incorporate advocacy planning at the neighborhood level into the planning process. It can address environmental justice by including justice values in planning education.

Figure 12.4 President Clinton signing the EJ executive order in 1994 [19].

- *Transportation:* Transportation is the largest source of pollution. The Department of Transportation (DOT) issued regulatory guidelines to address environmental justice in transportation in 1997. There appears to be discrimination in provision of transportation services in the eyes of advocacy group members. Hence, environmental justice is necessary in transportation planning and policy. The Department of Transportation ensures ES non-discrimination and considers adverse economic, social, and environmental effects of transportation. To be specific, DOT is committed to three basic principles of EJ: (1) Ensure full participation of low-income and minority groups and communities potentially affected by the transportation decision-making process; (2) Avoid, minimize, or mitigate disproportionately high and adverse human health and environmental effects on minority populations and low-income populations; and (3) Prevent the denial of, reduction in, or significant delay in the receipt of benefits by minority and low-income populations [20]. DOT should endeavor to reduce transport pollution.
- *Environmental Health:* Environmental justice should address socially-derived inequalities affecting health, well-being, and the environment. It must also account for the social, economic and political factors that influence the experiences of affected communities. Critical reviews have revealed substantial evidence that people living in poverty and certain racial and ethnic groups bear a disproportionate burden of environmental hazards. Environmental justice advocates contend that environmental health risks are borne disproportionately by members of the population who are poor and nonwhite. They question the extent to which society has achieved "equity" and "justice" in safeguarding the health and safety of its citizens. Environmental health research can contribute to fair and equitable protection for everyone, regardless of age, ethnicity, gender, race, or socioeconomic status.
- *Climate Justice:* This is perhaps the number one global environmental justice issue of the 21st century. Environmental injustice is key to understanding climate change. Scientific consensus indicates that climate change is a very real phenomenon caused by human activities.

Other areas of application of EJ include energy use, tourism, warehousing location, agricultural production, food security, and air pollution.

12.8 ENVIRONMENTAL JUSTICE IN DEVELOPED NATIONS

There is environmental justice distinction between developed countries and developing countries. Evidence shows that affluent countries are the primary drivers of global environmental degradation. As wealth concentrates in fewer hands, billions of people have to struggle to meet their basic necessities. The cases of the US, UK, Canada, Germany, and Australia are examined in detail.

- *United States*: The birth of the environmental justice movement in the US is often traced to the acts of civil disobedience by groups of predominantly African Americans protesting the dumping of toxic wastes in a landfill in Warren County, North Carolina. United States has made great strides in protecting the environment, while also ensuring fair treatment under the law. During the past decades, environmental justice has emerged as a socio-politically accepted term in the US, denouncing the unequal distribution of environmental burdens throughout the nation. Former manufacturing centers in North America and Europe have undergone deindustrialization, leading many toxic industries to relocate in the Global South [21]. Both Los Angeles and New York sought to plant a million trees, prioritizing underserved neighborhoods. Expanding the urban forest is a critical climate adaptation measure.
- *United Kingdom*: In England, the concept of environmental justice is seen to emerge at an elite civil society-level around the turn of the millennium, most noticeably with the Friends of the Earth (FoE), a non-government organization. The closest parallel to the EJ experience of US is the UK Black Environmental Network (BEN). The FoE plays a major role in environmental politics, governance, and promotion of EJ in UK [22].
- *Canada:* There is increasing evidence of environmental injustice in Canada, where individuals or communities experience disproportionate levels of harm from pollution. In contemporary environmental justice struggles across Canada, contests are being waged over the claims of expertise and counter-expertise. There is a growing tension between those who are allowed to have "knowledge" and those who are not. Individual liberty and the right to equality are grounded in Canadian Charter of Rights and Freedoms [23]. The charter is aimed at protecting the rights of all Canadians, especially the most vulnerable.
- *Germany:* In this country, environmental justice issues first emerged in debates on environmental health. The issues have gained more attention in the health sciences as issues encompassing the social environmental health inequalities. Socially disadvantaged people are more often exposed to outdoor air pollution than the better off and suffer more frequently from environmental diseases. The environmental justice has been introduced by a top down process, instead of being framed by a grassroots movement as in the United States. EJ is increasingly becoming a subject of academic debate in Germany. There are initiatives designed to promote environmental justice in Germany [24,25].
- *Australia:* This nation lacks easily identifiable environmental justice organizations or agencies, which are the key stakeholder groups in the design and implementation of environmental policies. Comparatively little has been written about environmental justice in Australia. Few Australian universities teach their students about environmental justice. Environmental injustices do exist in Australia, injustices which express inequality between socio-economic and socio-demographic groups [26].

12.9 ENVIRONMENTAL JUSTICE IN DEVELOPING NATIONS

Environmental justice organizations (EJOs) evolved in different nations. These EJOs are the Acción Ecológica, A Sud, the Center for Civil Society, Nature Kenya, Environmental Rights Action/Friends of the Earth Nigeria, and the World Rainforest Movement [27]. The underlying causes of environmental degradation, such as population growth, poverty, underdevelopment, inadequate technologies, and market failure due to market prices that do not take into account environmental impacts, will have to be addresses. A typical urban resilience in the Global South is shown in Figure 12.5 [28]. The cases of a range of developing nations such as China, India, Nigeria, South Africa, and Zambia are considered in detail.

Figure 12.5 A typical urban resilience in the Global South [28].

- *China:* Although inequality cannot be eliminated, it is necessary to provide equal opportunity to compete for all. The negative image of the People's Republic of China (PRC) as "the world's largest polluter" may be due to the fact that Chinese authorities are reluctant to address environmental issues and to take appropriate action. Environmental justice is mainly conceived as equity in access to environmental goods. Chinese environmental justice includes the principle of environmental rights, the principle of equal opportunity, fair competition for environmental resources, and the principle of compensation for vulnerable groups [29,30].
- *India:* This is a rapidly developing nation. Although the poor communities exist alongside the elite, no provision has been made to improve their situation. This often leads to some residents experiencing the terrible effects of environmental negligence. The Indian constitution does not specifically recognize a fundamental right to water. There is a link between the environment and water [31]. However, the Green Bench of the Supreme Court of India is hailed as the best example of response to environmental challenges. The Supreme Court mandated setting up specialized environmental courts. National Green Tribunal (NGT) was established in 2010 as an alternate forum to deliver speedy and inexpensive justice [32].

- *Nigeria:* Energy production is increasing, yet there is more inequality and energy poverty. The crude oil-induced violence and struggle of the Ogoni people in Niger delta is an effort about access to the revenues generated from the oil in their area. The Ogonis are indigenous communities in southeast Nigeria, rising to prominence in the environmental justice movement in the 1990s, under the leadership of activist Ken Saro-Wiwa. The Nigerian Constitution lacks the power that is needed in environmental protection. The Nigerian legal system should take their cue from other countries such as Ghana, South Africa, and India [33].
- *South Africa*: Environmental justice has developed considerably in post-apartheid South Africa. South Africa is bound by international environmental law. This implies that recognized principles of international environmental law such as sustainable development and EIA are part of the South African environmental law. The legislation establishes that environmental justice must be pursued so that adverse environmental impacts shall not be the case. The law also stipulates that there must be equitable access to environmental resources, benefits, and services. Resources should not be distributed in a manner that will unfairly discriminate against any person, and particularly vulnerable and disadvantaged persons. Ground Work is one of South Africa's environmental justice organizations and the local affiliate of Friend of the Earth [34,35].
- *Zambia:* This nation is classified as one of the world's least developed nations and is more likely to face the worst effects of global environmental challenges. The development and implementation of environmental law in Zambia is still in its infancy.

12.10 BENEFITS

Environmental justice/injustice is a call to equity in view of the disparities among people in costs and benefits distribution. It is an important issue affecting health disparities among citizens. It can contribute to fighting environmental racism, eliminating poverty. and reducing inequality. Environmental justice activists link the unequal distribution of resources with environmental racism. Economic equity is ensuring equitable distribution of economic benefits and costs among stakeholders, including residents [36]. Those experiencing the most harmful effects of a changing climate are typically those who have contributed the least emissions. For this reason, native Americans and other minority groups seek EJ restitution [37].

Environmental justice is useful for analyzing North-South environmental conflict as the North often dominates the decision-making process in international trade and financial institution. EJ promotes grassroots struggles that enable subordinated communities to speak for themselves and to challenge establishments and institutions. EJ activists seek to ask global questions about global problems. By so doing, they address concerns that mainstream organizations and government institutions traditionally ignore. GEJ may offer new opportunities for collaboration and exchange between engaged scholars and critical activists.

12.11 CHALLENGES

Environmental Justice Movement has not escaped criticism. There are challenges to collaboration between environmental justice advocates and the environmentalists who lead the climate change movement. Recognizing the plurality of the principles of environmental justice poses a challenge to those who seek to identify a set of universal principles of justice and sustainability. There is lack of standards and regulatory guidance in implementing environmental justice principles. Standardized measures are needed to inform public dialogue and policy [38]. There is also lack of environmental justice ethics in many nations.

Environmental justice is an important issue affecting health disparities since the environment could be regarded as a mediator between socioeconomic position and health-related outcomes. Modern environmentalism is no longer capable of dealing with the world's most serious ecological crisis. An effective response to the unprecedented global scale of the ongoing climate crisis demands a new kind of environmentalism. International environmental negotiations and agreements are marred by lack of adequate compliance and enforcement mechanisms. Until now, it has been impossible to properly document the prevalence and incidence of contentious activity related to environmental issues. These challenges can also produce new injustices or perpetuate existing ones.

12.12 CONCLUSION

Environmental justice embraces the principle that all people and communities are entitled to equal protection of our environmental laws. This concept originated from the civil rights campaigns of the 1960s and the more recent Environmental Justice Movement. The environmental justice movement has grown, matured, and taken root around the world. However, environmental injustice is still prevalent in the US and around the globe. Environmental injustice exists when members of disadvantaged ethnic minority or other groups suffer disproportionately. The fair treatment of all people is essential in creating a healthy world where children can grow to their full potential.

Environmental justice is a phenomenon that is causing revolution in the United States and abroad. It is based on the principle that all people have a right to be protected from environmental pollution. EJ scholarship has been expanding. It is increasingly becoming a major concern for activities and scholar worldwide. It is now a transnational and global issue.

One can only expect that the global push for environmental justice will keep growing. June 5th of every year has been declared World Environment Day. Citizens must be educated in organizing, mobilizing, and empowering themselves to take charge of their lives, their communities, and their environments. More information on environmental justice can be found in books in [9, 39-57] and several other books available online.

REFERENCES

[1] "The world's top 10 battles for environmental justice,"

https://cosmosmagazine.com/geoscience/the-world-s-top-10-battles-for-environmental-justice

[2] M. N. O. Sadiku, O. D. Olaleye, and S. M. Musa, "Environmental justice: A primer," *International Journal of Trend in Research and Development*, vol. 6, no. 5, Sept.-Oct. 2019, pp. 31-34.

[3] C. G. Gonzalez and S. Atapattu, "International environmental law, environmental justice, and the global South," *Transnational Law & Contemporary Problems*, vol. 26, no. 2, p. 229, June 2017, pp. 229-242.

[4] Q. Yua, "Environmental justice in warehousing location: State of the art," *Journal of Planning Literature*, vol. 33, no. 3, 2018, pp. 287-298.

[5] "The American Environmental Justice Movement,"

https://iep.utm.edu/enviro-j/

[6] "Wisconsin bail out the People Movement,"

https://wibailoutpeople.org/2016/02/22/peoples-power-assembly-in-solidarity-with-flint/

[7] M. L. Jaramillo, *Almost Everything You Need to Know About Environmental Justice,*

http://d3n8a8pro7vhmx.cloudfront.net/unitedchurchofchrist/legacy_url/421/almost-everything-you-need-to-know-about-environmental-justice-english-version.pdf?1418423801

[8] "Principles of environmental justice,"

https://www.ejnet.org/ej/principles.html

[9] A. Park, *Everybody's Movement: Environmental Justice ad Climate Change.* Washington DC: Environmental Support Center, 2009.

[10] J. Mousie, "Global environmental justice and postcolonial critique," *Environmental Philosophy*, vol. 9, no. 2, Fall 2012, pp. 21-46.

[11] "Environmental justice," *Wikipedia*, the free encyclopedia,

https://en.wikipedia.org/wiki/Environmental_justice

[12] S. Chamber, "Minority empowerment and environmental justice," *Urban Affairs Review,* vol. 43, no. 1, September 2007, pp. 28-54.

[13] "Environmental justice everywhere," June 2019,

https://paxchristiusa.org/2019/06/12/webinar-replay-environmental-justice-everywhere/

[14] J. Martinez-Alier et al., "Is there a global environmental justice movement?" *The Journal of Peasant Studies,* vol. 43, no. 3, 2016, pp. 731-755.

[15] M. N. O. Sadiku, O. D. Olaleye, and S. M. Musa, "Global environmental justice," *International Journal of Trend in Research and Development*, vol. 6, no. 5, October 2019, pp. 55-59.

[16] "Global atlas of 'environmental justice' re-launches website tracking conflicts involving companies,"

https://www.business-humanrights.org/en/global-atlas-of-environmental-justice-re-launches-website-tracking-conflicts-involving-companies

[17] T. Sikor and P. Newell, "Globalizing environmental justice?" *Geoforum,* vol. 54, July 2014, pp. 151-157.

[18] D. C. Esty and M. H. Ivanova, "Global environmental governance: Options & opportunities," *Forestry & Environmental Studies Publications Series.* 8, 2002.

[19] "Learn about environmental justice,"

https://www.epa.gov/environmentaljustice/learn-about-environmental-justice

[20] S. Sen, "Environmental justice in transportation planning and policy: A view from practitioners and other stakeholders in the Baltimore-Washington, D.C. metropolitan region," *Journal of Urban Technology,* vol. 15, no. 1, 2008, pp. 117–138.

[21] R. Schroeder et al., "Third World Environmental Justice," *Society and Natural Resources*, vol. 21, no. 7, 2008, 547-555,

[22] D. Stern snf W.Wolfe (2016) "The lost war and battles of environmental justice: The emergence of environmental justice in England political potential in a post-political context," *Masters Theses,* Durham University. Available Online: http://etheses.dur.ac.uk/11900/

[23] D. N. Scott, " 'We are the monitors now': Experiential knowledge, transcorporeality and environmental justice," *Social & Legal Studies*, vol. 25, no. 3, 2016, pp. 261–287.

[24] H. D. Elvers, M. Gross, and H. Heinrichs, "The diversity of environmental justice," *European Societies*, vol.10, no. 5, 2008, pp. 835-856.

[25] L. Strelau and H. Köckler (2016) "'It's optional, not mandatory': Environmental justice in local environmental agencies in German," *Local Environment,* vol. 21, no. 10, 2016, pp. 1215-1229.

[26] J. Byrne and D. MacCallum, "Bordering on neglect: Environmental justice in Australian planning," *Australian Planner*, vol. 50, no. 2, 2013, pp.164-173.

[27] B. Rodriguez- Labajos et al., "Not so natural an alliance? degrowth and environmental justice movements in the global South," *Ecological Economics*, vol. 157, March 2019, pp. 175-184.

[28] A. Allen, L. Griffin, and C. Johnson, *Environmental Justice and Urban Resilience in the Global South.* Palgrave. 2017.

[29] L. Lanhai and Z. Hongjing, "China green logistics environmental justice

ethics construction," *World Automation Congress,* June 2012.

[30] R. Balme, "Mobilising for environmental justice in China, Asia Pacific," *Journal of Public Administration,* vol. 36, no. 3, 2014, pp. 173-184.

[31] L. Mehta et al., "Global environmental justice and the right to water: The case of peri-urban Cochabamba and Delhi," *Geoforum,* vol. 54, July 2014, pp. 158-166.

[32] S. Shrotria, "Environmental justice: Is the National Green Tribunal

of India effective?" *Environmental Law Review,* vol. 17, no. 3, 2015, pp. 169–188.

[33] K. I. Ajibo, "Transboundary hazardous wastes and environmental justice: Implications for economically developing countries," *Environmental Law Review,* vol. 18, no. 4, 2016, pp. 267–283.

[34] P. T. Sambo, "A conceptual analysis of environmental justice approaches: Procedural environmental justice in the EIA process in South Africa and Zambia, *Doctoral Dissertation*, University of Manchester, 2012.

[35] B. Peek, "groundWork environmental justice action climate change letter to South African President Cyril Ramaphosa, December 2018," *NEW SOLUTIONS: A Journal of Environmental and Occupational Health Policy*, vol. 29, no. 1, 2019, pp. 112–115.

[36] S. Lee and T. Jamal, "Environmental justice and environmental equity in tourism: Missing links to sustainability," *Journal of Ecotourism*, vol. 7, no. 1, 2008, pp. 44-67.

[37] J. Vickery and L. M. Hunter, "Native Americans: Where in environmental justice research?" *Society and Natural Resources*, vol. 29, vol. 1, 2006, pp, 36-52.

[38] J. Harner et al. "Urban environmental justice indices," *The Professional Geographer*, vol. 54, no. 3, 2002, pp. 318-331.

[39] B. Bryant (ed.), *Environmental Justice: Issues, Policies, and Solutions*. Washington DC: Island Press, 1993.

[40] J. Agyeman, *Sustainable Communities and the Challenge of Environmental Justice. New* York: New York University Press, 2005.

[41] D. V. Carruthers (ed.), *Environmental Justice in Latin America: Problems, Promise, and Practice.* MIT Press, 2008.

[42] D. N. Pellow, *What is Critical Environmental Justice? Polity.* Cambridge, 2018.

[43] L. Dominelli, *Green Social Work: From Environmental Crises to Environmental Justice*. Cambridge, 2012.

[44] T. Shallcross and J. Robinson (eds.), *Global Citizenship and Environmental Justice*. Editions Rodopi, 2006.

[45] D. E. Clover (ed.), *Global Perspectives in Environmental and Adult Educatio, Justice, Sustainability, and Transformation.* Peter Lang Publishing Inc., 2003.

[46] C. G. Boone and M. Fragkias (eds.), *Urbanization and Sustainability: Linking Urban Ecology, Environmental Justice And Global Environmental Change*. Springer, 2013.

[47] F. C. Steady, *Environmental Justice in the New Millennium: Global Perspectives on Race, Ethnicity, and Human Rights.* New York: Palgrave Macmillan, 2009.

[48] D. N. Pellow, *Resisting Global Toxics: Transnational Movements for Environmental Justice.* Cambridge: MIT Press, 2007.

[49] C. Okereke, *Global Justice and Neoliberal Environmental Governance: Ethics, Sustainable Development and International Co-operation.* New York: Routledge, 2007.

[50] L. Leonard and S. B. Kedzior (eds.), *Occupy the Earth: Global Environmental Movements.* Emerald Group Publishing Limited, 2014.

[51] J. D. Wulfhorst and A.K. Haugestad (eds.), *Building Sustainable Communities; Environmental Justice & Global Citizenship.* Editions Rodopi, 2006.

[52] J. Lester, D. Allen, and K. M. Hill, *Environmental Injustice in the U.S.:Myths and Realities.* Routledge, 2000.

[53] D. Schlosberg, *Defining Environmental Justice: Theories, Movements, and Nature.* Oxford University Press, 2007.

[54] P. S. Wenz, *Environmental Justice.* Albany, NY: State University of New York Press, 1988.

[55] L. W. Cole and S. R. Foster, *From the Ground Up: Environmental Racism and the Rise of the Environmental Justice Movement.* New York: University Press, 2001.

[56] G. Walker, *Environmental Justice: Concepts, Evidence and Politics.* London, UK: Routledge, 2012.

[57] K. Shrader-Frechette, *Environmental Justice: Creating Equality, Reclaiming Democracy.* Oxford University Press, 2002.

APPENDIX A

OTHER TOPICS

This appendix covers other topics on environmental studies not covered as chapters in the book or not covered at all.

Ecological Restoration
Environmental Accounting
Environmental Anthropology
Environmental Assessment
Environmental Attitudes
Environmental Awareness
Environmental Bias
Environmental Biology
Environmental Chemistry
Environmental Conservation
Environmental Construction
Environmental Crime
Environmental Damage
Environmental Degradation
Environmental Democracy
Environmental Design
Environmental Economics
Environmental Effects on physiology
Environmental Enterprises
Environmental Epidemiology
Environmental Ethics
Environmental Ethics and Legislation
Environmental Facilities
Environmental Federalism
Environmental Finance
Environmental Geography
Environmental Geology
Environmental Governance
Environmental History
Environmental Ignorance
Environmental Impact assessment
Environmental Information

Environmental Integrity
Environmental Issues
Environmental Knowledge
Environmental Leadership
Environmental Legislation
Environmental Literacy
Environmental Lobbyist
Environmental Magnetism
Environmental Management
Environmental Medicine/Healthcare
Environmental Monitoring
Environmental Movement
Environmental Nanotechnology
Environmental Organization
Environmental Philosophy
Environmental Physics
Environmental Planning
Environmental Politics
Environmental Policy Analyst
Environmental Practice
Environmental Protection
Environmental Psychology
Environmental Quality
Environmental Racism
Environmental Remediation
Environmental Restoration
Environmental Sexism
Environmental Sociology
Environmental Soil science
Environmental Stewardship
Environmental Sustainability
Environmental Technology
Environmental Toxicology
Environmental Values
Environmentalism

APPENDIX B

SELECTED BIBLIOGRAPHY

Abate, R. S., *Directory of Environmental Law: Education Opportunities at American Law Schools.* Carolina Academic Press, 2nd edition, 2008.

Abbasi, S. A. and T. Abbasi, *Current Concerns in Environmental Engineering.* Nova Science Publishers, 2018.

Adams, E., J. Bartram, and Y. Chartier (eds.), *Essential Environmental Health Standards in Health Care.* World Health Organization, 2008.

Addams, H., J. L. R. Proops (eds.), *Social Discourse and Environmental Policy: An Application of Q Methodology.* Edward Elgar, 2000.

Adelle C., K. Biedenkopf, and D. Torney (eds.), *European Union External Environmental Policy: Rules, Regulation and Governance Beyond Borders.* Palmgrave Macmillan, 2018.

Agyeman, J., *Sustainable Communities and the Challenge of Environmental Justice.* New York: New York University Press, 2005.

Ahluwalia, V. K., *Advanced Environmental Chemistry.* The Energy and Resources Institute (TERI), 2017.

Ali, S. Z., *Environmental Biology.* Akhand Publishing House, 2019.

Allaby, M., *Basics of Environmental Science.* Routledge, 2nd edition, 2000.

Allan C. and G. H. Stankey, *Adaptive Environmental Management.* Springer, 2009.

Anderson, D. A. *Environmental Economics and Natural Resource Management.* London, UK: Routledge, 4th edition, 2013.

Anderson, F. R. *NEPA in the Courts: A Legal Analysis of the National Environmental Policy Act.* Taylor & Francis, 2013.

Andrews, J. E. et al., *An Introduction to Environmental Chemistry.* Blackwell Science Ltd, 2nd edition, 2004.

Andrews, R. N. L., *Managing the Environment, Managing Ourselves: A History of American Environmental Policy.* Yale University Press, 2nd edition, 2006.

Aravossis, K. et al. (eds.), *Environmental Economics Environmental Economics and Policy.* WIT Press, 2006.

Asafu-Adjaye, J., *Environmental Economics for Non-Economists: Techniques and Policies for Sustainable Development.* World Scientific, 2nd Edition, 2005.

Atapattu, S., *Emerging Principles of International Environmental Law.* Brill, 2007.

Atapattu, S., *Emerging Principles of International Environmental Law.* Transnational Publishers, 2006.

Bailer A. J., *Statistics for Environmental Biology and Toxicology.* Boca Raton, FL: CRC Press, 2020.

Barrow, C., *Environmental Management for Sustainable Development.* New York: Routledge, 2nd edition, 2006.

Bassett, W. H., *Clay's Handbook of Environmental Health.* London, UK: Son Press, 19th edition, 2004.

Bateman, I. J., A. A. Lovett, and J. S. Brainard, *Applied Environmental Economics*: A GIS Approach to Cost-Benefit Analysis. Academic, 2005.

Baumol, W. J. and W. E. Oates, *The Theory of Environmental Policy.* Cambridge University Press, 2nd edition, 2012.

Bell, C. et al., *Environmental Law Handbook.* Bernan Press, 23rd edition,2016.

Benelmir, R. (ed.), *Energy-Environment-Economics.* Nova Science Publishers, 2014. Reading, MA: Addison Wesley, 2000.

Berkhout, F., M. Leach, and I. Scoones, *Negotiating Environmental Change: New Perspectives from Social Science.* Edward Eglar, 2003.

Berkhout, F., *Negotiating Environmental Change: New Perspectives from Social Science.* Edward Elgar, 2003.

Best, G., *Environmental Pollution Studies.* Liverpool University Press, 2000.

Betts, K., *A Survey of Environmental Chemistry Around the World: Studies, Processes, Techniques, and Employment.* American Chemical Society, 2014.

Betts, K., *A Survey of Environmental Chemistry Around the World: Studies, Processes, Techniques, and Employment.* American Chemical Society, 2014.

Beyerlin, U. and T. Marauhn, *International Environmental Law.* Oxford, UK: Hart Publishing, 2011.

Bhakta, J. N., S. Lahiri and B. B. Jana (eds.), *Green Technology for Bioremediation of Environmental Pollution.* Nova Science Publishers, 2019.

Bhakta, J. N., S. Lahiri and B. B. Jana (eds.), *Green Technology for Bioremediation of Environmental Pollution.* Nova Science Publishers, 2019.

Bhatt, M. S., S. Ashraf, and A. Illiyan (eds.), *Problems and Prospects of Environment Policy: Indian Perspective.* Aakar Books, 2008.

Bodansky, D., J. Brunnée, and E. Hey, *The Oxford Handbook of International Environmental Law.* Oxford, 2008.

Bodansky, D., *The Art and Craft of International Law.* Cambridge, MA: Harvard University Press, 2010.

Bodin, Ö. and C. Prell, *Social Networks and Natural Resource Management: Uncovering the Social Fabric of Environmental Governance.* Cambridge, UK: Cambridge University Press, 2011.

Bodzin, A., B. S. Klein, and S. Weaver (eds.), *The Inclusion of Environmental Education in Science Teacher Education.* Springer, 2010.

Boeker, E., and R. V, Grondelle, *Environmental Physics: Sustainable Energy and Climate Change.* John Wiley & Sons, 4th edition, 2013.

Boer, B., R. Ramsay, and D. R. Rothwell, *International Environmental Law in the Asia Pacific.* Kluwer Law International, 2006.

Boone C. G. and M. Fragkias (eds.), *Urbanization and Sustainability: Linking Urban Ecology, Environmental Justice and Global Environmental Change.* Springer, 2013.

Borghese, F., P. Denti, and R. Saija, *Scattering from Model Nonspherical Particles: Theory and Applications to Environmental Physics*. Springer, 2007.

Borzsák, L., *The Impact of Environmental Concerns on The Public Enforcement Mechanism Under EU Law: Environmental Protection in the 25th Hour*. Wolters Kluwer Law & Business, 2011.

Botkin, D. B. and E. A. Keller, *Environmental Science: Earth as a Living Planet*. John Wiley & Sons, 2002.

Boyd, D. R., *Unnatural Law: Rethinking Canadian Environmental Law and Policy*. Vancouver, Canada: The University of British Columbia Press, 2003.

Brady, J., A. Ebbage, and R. Lunn, *Environmental Management in Organizations: The IEMA Handbook*. London, Routledge, 2nd Edition, 2011.

Brinkman, A. W., *Physics of the Environment*. Imperial College Press, 2008.

Brown, P. and L. Gibbs, *Toxic Exposures: Contested Illnesses and the Environmental Health Movement*. Columbia University Press, 2007.

Buchholz, W. and D. Rübbelke, *Foundations of Environmental Economics*. Springer International Publishing, 2019.

Bullard, R. D., G. S. Johnson, and A. O. Torres, *Environmental Health and Racial Equity in the United States: Building Environmentally Just, Sustainable, and Livable Communities*. Amer Public Health Association, 2011.

Buller, H., G. A. Wilson, and A. Holl, *Agri-Environmental Policy in the European Union*. Ashgate Publishing Ltd., 2000.

Burgman, M., *Risks and Decisions for Conservation and Environmental Management*. Cambridge, UK: Cambridge University Press, 2005.

Burke, A., *Development and Environmental Policy in India*. CreateSpace Independent Publishing. 2018.

Callan, S. and J. Thomas, *Environmental Economics and Management: Theory, Policy and Applications*. Fort Worth: Dryden Press, 2nd edition, 2000.

Calvin W. Rose, *An Introduction to the Environmental Physics of Soil, Water and Watersheds*. Cambridge University Press, 2012.

Camilleri, M. A., *Corporate Sustainability, Social Responsibility and Environmental Management: An Introduction to Theory and Practice with Case Studies*. Springer, 2017.

Carraro C. and F. Lévêque, *Voluntary Approaches in Environmental Policy: An Assessment*. OECD Publishing, 2013.

Carraro, C., Y. Katsoulacos, and A. Xepapadeas, *Environmental Policy and Market Structure*. Springer Science & Business Media, 2013.

Carruthers D. V. (ed.), *Environmental Justice in Latin America: Problems, Promise, and Practice*. MIT Press, 2008.

Chadwick A. et al., *Hydraulics in Civil and Environmental Engineering*. London, CRC Press, 5th ed., 2013.

Chadwick, A., J. Morfett, and M. Borthwick, *Hydraulics in Civil and Environmental Engineering*. Boca Raton, FL: CRC Press, 5th edition, 2010.

Chapman, D., *Environmental Economics - Theory, Application, and Policy*. Reading, MA: Addison Wesley, 2000.

Chauhan, A., *Environmental Pollution and Management.* Delhi, India: I.K. International Publishing House, 2019.

Clover D. E. (ed.), *Global Perspectives in Environmental and Adult Education, Justice, Sustainability, and Transformation.* Peter Lang Publishing Inc., 2003.

Cohen, S., *Understanding Environmental Policy.* Columbia University Press, 2006.

Cole, L. W. and S. R. Foster, *From the Ground Up: Environmental Racism and the Rise of the Environmental Justice Movement.* New York: University Press, 2001.

Corbitt, R. A., *Standard Handbook of Environmental Engineering.* McGraw-Hill, 2nd ed., 2004.

Coyle, S. and K. Morrow, *The Philosophical Foundations of Environmental Law.* Property, Rights and Nature. Hart Publishing 2004.

Cullet P., *Differential Treatment in International Environmental Law.* Ashgate Publishing Co., 2003.

Cunningham, W. and M. A. Cunningham, *Principles of Environmental Science.* McGraw-Hill Education, 9th edition, 2019.

Cunningham, W. P., M. A. Cunningham, and B. W. Saigo, *Environmental Science: A Global Concern.* McGraw-Hill Education, 15th edition, 2020.

DaS, T. K., *Industrial Environmental Management: Engineering, Science, and Policy.* John Wiley & Sons, 2020.

Dasgupta, P., S K. Pattanayak, and V. Kerry Smith (eds.), *Handbook of Environmental Economics.* North Holland, Volume 4, 2018.

Davies, P. G. G., *European Union Environmental Law: An Introduction to Key Selected Issues.* Ashgate, 2004.

Davis, M. and S. Masten, *Principles of Environmental Engineering and Science.* McGraw-Hill, 3rd ed, 2013.

Davis, M. L. and D. A. Cornwell, *Introduction to Environmental Engineering.* McGraw-Hill, 5th ed., 2013.

de Sadeleer, N., *EU Environmental Law and The Internal Market.* Oxford University Press, 2014.

Decision Making for the Environment: Social and Behavioral Science Research Priorities. National Academies Press, 2005.

Deeba, F., *Ecology and Environmental Biology.* Centrum Press, 2017.

Delreux, T. and S. Happaerts, *Environmental Policy and Politics in the European Union.* Red Globe Press, 2016.

Desai, B. H., *Institutionalizing International Environmental Law.* Transnational Publishers, 2004.

Dominelli, L., *Green Social Work: From Environmental Crises to Environmental Justice.* Cambridge, 2012.

Dupuy, P. M. and J. E. Vinuales, *International Environment Law.* Cambridge University Press, 2nd edition, 2018.

Duxbury, R. and S. Morton (eds.), *Blackstone's Statutes on Environmental Law.*

Oxford University Press, 5th edition, 2004.

Ebbeson, J. and P. Okowa (eds.), *Environmental Law and Justice in Context.* Cambridge, UK: Cambridge University Press, 2009.

Eccleston C. H. and F. March, *Global Environmental Policy: Concepts, Principles, and Practice.* Boca Raton, FL: CRC Press, 2011.

Encyclopaedia of Introduction to Environmental Physics: Planet Earth, Life and Climate (4 Volumes). Hillingdon, UK: Publisher Koros Press Limited, 2015.

Encyclopedia of Environmental Health. Elsevier, 5-Volumes, 2nd edition, 2019.

Enger, E. D., B. F. Smith, and A. T. Bockarie, *Environmental Science: A Study of Interrelationships.* McGraw-Hill Education, 15th edition, 2018.

Etty, T. F. M. et al., (eds.), *The Yearbook of European Environmental Law.* Oxford University Press, 2005.

Faraoni, V., *Exercises in Environmental Physics.* Springer, 2006

Faure, M. and G. Heine, *Criminal Enforcement of Environmental Law in the European Union.* Kluwer Law, 2005.

Field B. and M. K. Field, *Environmental Economics: An Introduction.* McGraw-Hill, 7th Edition. 2017.

Fifield, F. W. and W. P. J. Hairens, *Environmental Analytical Chemistry.* Black Well Science Ltd, 2nd edition, 2000.

Finn, S. and L. R. O'Fallon (eds.), *Environmental Health Literacy.* Springer, 2019.

Fisher, E., *Environmental Law - A Very Short Introduction.* Oxford University Press, 2018.

Fisher, M. R., *Environmental Biology.* Open Oregon Educational Resources, 2018.

Fitzgerald, P. L., *International Issues in Animal Law: The Impact of International Environmental and Economic Law Upon Animal Interests and Advocacy.* Carolina Academic Press, 2012.

Fitzmaurice, M., *Contemporary Issues in International Environmental Law.* Cheltenham, Edward Elgar, 2009.

Forinash, K., *Foundations of Environmental Physics: Understanding Energy Use and Human Impacts.* Washington DC: Island Press, 2010.

Forsyth, T., *Critical Political Ecology: The Politics of Environmental Science.* Taylor and Francis, 2003.

Franzie, S., B. Markert, and S. Wunschmann, *Introduction to Environmental Engineering.* Weinhei, Germany: Wiley-VCH Verlag, 2012.

Friis, R. H., *Essentials of Environmental Health (Essential Public Health).* Jones & Bartlett Learning, 3rd edition, 2018.

Frumkin, H., *Environmental Health: From Global to Local (Public Health/Environmental Health).* Jossey-Bass, 3rd edition, 2016.

Galanakis, C. M., *Innovation Strategies in Environmental Science.* Elsevier, 2019.

Ganesamurthy, V. S., *Environmental Economics in India.* New Delhi, India: New Century Pub., 2009.

Gaur, R. C., *Basic Environmental Engineering.* New Delhi, India: New Age International Publishers, 2008.

Gaurina-Medjimurec, N. (ed.), *Handbook of Research on Advancements in Environmental Engineering.* Engineering Science Reference, 2015.

Gillespie, A. *International Environmental Law, Policy, and Ethics.* Oxford University Press, 2nd edition, 2014.

Gilpin, A., *Environmental Economics: A Critical Overview.* John Wiley & Sons, 2000.

Goldfein, M. D. and A. V. Ivanov, *Applied Natural Science: Environmental Issues and Global Perspectives.* Apple Academic Press, 2016.

Goldstein, I. F. and M. Goldstein, *How Much Risk? A Guide to Understanding Environmental Health Hazards.* New York, NY: Oxford University Press, 2002.

Gray, M., J. Coates, and T. Hetherington (eds.), *Environmental Social Work.* Routledge, 2013.

Greenberg, M. R., *Environmental Policy Analysis and Practice.* Rutgers University Press, 2008.

Gupta, M., *Fundamentals of Environmental Biology.* I K International Publishing House, 2018.

Halvorsen, K. E. et al. (eds.), *A Research Agenda for Environmental Management.* Edwards Elgar Publishing, 2019.

Hanley, N., J. F. Shogren, and B. White, *Environmental Economics in Theory and Practice.* Palgrave Macmillan, 2nd edition, 2007.

Hannigan, J., *Environmental Sociology.* New York: Routledge, 2nd edition, 2006.

Hanrahan, G., *Key Concepts in Environmental Chemistry.* Waltham, MA: Academic Press, 2012.

He, Z. (ed.), *Environmental Chemistry of Animal Manure.* Nova, 2011.

Heinsohn, R. J. and J. M. Cimbala, *Indoor Air Quality Engineering: Environmental Health and Control of Indoor Pollutants.* Boca Raton, FL: CRC Press, 2013.

Hellawell, J. M., *Biological Indicators of Freshwater Pollution and Environmental Management.* Elsevier Science Publishers, 2012.

Herson, A. I. and G. A. Lucks, *California Environmental Law & Policy.* Solano Press Books; 2nd edition, 2017.

Hessing, M. and T. Summerville, *Canadian Natural Resource and Environmental Policy: Political Economy and Public Policy.* UBC Press, 2nd edition, 2014.

Hill, M. K., *Understanding Environmental Pollution.* Cambridge, UK: Cambridge University Press, 3rd edition, 2020.

Hillel, D. *Environmental Soil Physics.* San Diego, CA: Academic Press, 2003.

Hillel, D., *Environmental Soil Physics: Fundamentals, Applications, and Environmental Considerations.* Academic Press, 2004.

Hites, R. A., *Elements of Environmental Chemistry.* John Wiley & Sons, 2007.

Hughes, P. and N.J. Mason, *Introduction to Environmental Physics: Planet Earth, Life and Climate.* London, UK: CRC Press, 2014.

Hussen, A. M., *Principles of Environmental Economics.* Routledge, 2nd edition, 2004.

Hynes, H. P. and D. Brugge, *Community Research in Environmental Health: Studies in Science, Advocacy and Ethics.* Routledge, 2005.

Ibanez, J. G. et al., *Environmental Chemistry Fundamentals.* Springer, 2007.

Ibanez, J. G. et al., *Environmental Chemistry: Microscale Laboratory Experiments.* Springer, 2008.

International Environmental Law: Multilateral Environmental Agreements. Hanoi, International Publishing House, May 2017.

Jaeger, W. K., *Environmental Economics for Tree Huggers and Other Skeptics.* Island Press, 2005.

Jans, J. H. and H. H. B. Vedder, *European Environmental Law: After Lisbon.* Europa Law Publishing, 4th edition, 2012.

Jans, J. H., *European Environmental Law.* Europa Law Publishing, 2nd edition, 2000.

Jones, J. A. A., *Global Hydrology: Processes, Resources and Environmental Management.* Routledge, 2014.

Jordan A. and C. Adelle, *Environmental Policy in the EU: Actors, Institutions and Process.* Routledge, 3rd edition, 2013.

Jordan, A. and A. Lenschow (eds.), *Innovation in Environmental Policy? Integrating the Environment for Sustainability.* Edward Eglar Publishing, 2020.

Jordan, A., *Environmental Policy in the European Union.* Routledge, 2005.

Kaiser, J., J. E. Klanning, and L. E. Erickson, *Bioindicators and Biomarkers of Environmental Pollution and Risk Assessment.* Science Publishers, 2001.

Karpagam, M., *Environmental Economics A Textbook.* Sterling Publishers, 2019.

Kaur, A., Environment Education. Twenty First Century, 2017.

Keen, M., V. A. Brown, and R. Dyball (eds.), *Social Learning in Environmental Management: Towards a Sustainable Future.* Earthscan, 2005.

Khoiyangbam, R. S. and N. Gupta, *Introduction to Environmental Sciences.* New Delhi, India: The Energy and Resources Institute, 2012.

Khopkar, S. M., *Environmental Pollution Monitoring and Control.* New Delhi, India: New Age International Limited Publishers, 2007.

Kiss A. and D. Shelton, *Guide to International Environmental Law.* Martinus Nijhoff Publishers, 2007.

Kneese A. V. and B. T. Bower, *Environmental Quality Analysis: Theory & Method in The Social Sciences.* The Johns Hopkins University Press, 2013.

Knill C. and A. Lenschow, *Implementing EU Environmental Policy: New Directions and Old Problems.* Manchester University Press, 2000.

Knill C. and D. Liefferink, *Environmental Politics in the European Union: Policy-Making, Implementation and Patterns of Multi-Level Governance.* Manchester University Press, 2007.

Knoepfel, P., *Environmental Policy Analyses: Learning from the Past for the Future - 25 Years of Research.* Springer 2007.

Koivurova, T., *Introduction to International Environmental Law.* London, UK: Routledge, 2014.

Kolstad, C., *Intermediate Environmental Economics: International Edition.* India: Oxford University Press, 2nd edition, 2011.

Konisky D. M. (ed.), *Handbook of U.S. Environmental Policy.* Edward Elgar Publishing, 2020.

Koren, H. and M. Bisesi, *Handbook of Environmental Health and Safety, Principles and Practices.* Boca Raton, FL: CRC Press, volume 1, 4th ed., 2002.

Koren, H., *Best Practices for Environmental Health: Environmental Pollution, Protection, Quality and Sustainability.* Routledge, 2017.

Kotzé, L. J., *Global Environmental Governance; Law and Regulation for the 21st Century*. Edward Elgar Publishing, 2012.

Kraft, M. E. and Kamieniecki (eds.), *The Oxford Handbook of U. S. Environmental Policy.* Oxford University Press, 2013.

Kraft, M. E., *Environmental Policy and Politics*. Routledge, 6th edition, 2017

Krämer, L. (ed.), *European Environmental Law.* Ashgate Publishing Co., 2003.

Krishna, I. V. M., V. Manickam, and A. Shah, *Environmental Management: Science and Engineering for Industry.* India: BS Publications, 2017.

Kumar, M. and R. R. Tiwari (eds.), *Recent Trends and Advances in Environmental Health.* Nova Science Publishers, 2019.

Kuo, J. (ed.), *Air Pollution Control Engineering for Environmental Engineers*. Boca Raton, FL: CRC Press, 2016.

Kutz, M. (ed.), *Handbook of Environmental Engineering*. John Wiley & Sons, 2018.

Lahiri, S. (ed.), *Environmental Education*. New Delhi, India, Studera Press, 2019.

Lazarus, R. J., *The Making of Environmental Law.* Chicago: The University of Chicago Press, 2004.

Leary D. and B. Pisupati (eds.), *The Future of International Environmental Law*. Tokyo, Japan: United Nations University, 2010

Lee, C. C., *Environmental Engineering Dictionary*. Anham, 4th edition, 2005

Lee, C. C., *Handbook of Environmental Engineering Calculations*. McGraw-Hill, 2000

Lee, M., *EU Environmental Law, Governance and Decision Making*. Hart, 2014.

Lehmann, J. and S. Joseph, *Biochar for Environmental Management: Science, Technology and Implementation*. New York: Routledge, 2015.

Leonard, L. and S. B. Kedzior (eds.), *Occupy the Earth: Global Environmental Movements*. Emerald Group Publishing Limited, 2014.

Leroy, P., and A. Crabb, *The Handbook of Environmental Policy Evaluation*. Earthscan, 2012.

Lester, J., D. Allen, and K. M. Hill, *Environmental Injustice in the U.S.: Myths and Realities*. Routledge, 2000.

Levy B. S. et al. (eds.), *Occupational and Environmental Health: Recognizing and Preventing Disease and Injury.* Oxford University Press, 6th edition, 2011.

Lin B. C. and S. Zheng (eds.), *Environmental Economics and Sustainability*. John Wiley & Sons, 2017.

Lloro-Bidart, T. and V. Banschbach (eds.), *Animals in Environmental Education: Interdisciplinary Approaches to Curriculum and Pedagogy*. Palgrave Macmillan, 2019.

Lockyer, J. and J. R. Veteto, *Environmental Anthropology Engaging Ecotopia: Bioregionalism, Permaculture, and Ecovillages*. Berghahn Books, 2013.

Louka, E., *International Environmental Law: Fairness, Effectiveness and World Order*. Cambridge University Press,2006.

Lynn, L. M., *Environmental Biology and Ecology Laboratory Manual*. Kendall Hunt Publishing; 6th edition, 2016.

Mackay D. and R. S. Boethling, *Handbook of Property Estimation Methods for Chemicals: Environmental Health Sciences*. Boca Raton, FL: CRC Press, 2000.

Mäler, K. G. and J. R. Vincent (eds.), *Handbook of Environmental Economics: Valuing Environmental Changes*. Amsterdam: Elsevier/North-Holland, Volume 2, 2005.

Malone, L. A., *Environmental Law*. Aspen Publishers, 2003.

Managi S. and K. Kuriyama, *Environmental Economics*. Routledge, 2016.

Managi, S. (ed.), *The Routledge Handbook of Environmental Economics in Asia*. Routledge, 2019.

Managi, S. and S. Kaneko, *Chinese Economic Development and the Environment:*

New Horizons in Environmental Economics Series. Cheltenham, UK: Edward Elgar. 2009.

Manahan, S. E., *Environmental Chemistry*. Boca Raton, FL: CRC Press, 10the edition, 2017.

Manly, B. F. J., *Statistics for Environmental Science and Management*. Boca Raton, FL: CRC Press, 2008.

Markandya A. et al., *Environmental Economics for Sustainable Growth: A Handbook for Practitioners*. Cheltenham, UK: Edward Elgar, 2002.

Maskall, J. and A. Stokes, *Designing Effective Fieldwork for the Environmental and Natural Sciences*. Plymouth: The Higher Education Academy Subject Centre for Geography, 2008.

Mason, N. and P. Hughes, *Introduction to Environmental Physics: Planet Earth, Life and Climate*. Taylor and Francis, 2001.

Masters, G. M. and W. P. Ela, *Introduction to Environmental Engineering and Science*. Pearson, 3rd edition, 2007.

Masters, G. M. and W. P. Ela, *Introduction to Environmental Engineering and Science*. Pearson Education Limited, 2013.

Maxwell, N. I., *Understanding Environmental Health: How We Live in the World*. Burlington, MA: Jones and Bartlett Publishers, 2nd edition, 2014.

Maxwell, N. I., *Understanding Environmental Health: How We Live in the World*. Jones & Bartlett Learning, 2nd edition, 2013.

Mazmanian D. A. and M. E. Kraft (eds.), *Toward Sustainable Communities: Transition and Transformations in Environmental Policy*. Cambridge, MA: MIT Press, 2009.

McBeath, G. A. et al., *Environmental Education in China*. Edward Elgar Publishing, 2015.

McCally, M., *Life Support: The Environment and Human Health*. The MIT Press, 2002.

McEldowney, J. F. and S. McEldowney, *Environmental Law and Regulation*. Blackstone Press, 2001.

McKinney, M. L. et al., *Environmental Science: Systems and Solutions*. Sudbury, MA: Jones and Barlett Publishers, 4th edition, 2007

McKinney, R. E., *Environmental Pollution Control Microbiology: A Fifty-Year Perspective*. Boca Raton, FL: CRC Press, 2004.

McManus, F., *Environmental Law in Scotland: An Introduction and Guide*. Edinburgh University Press, 2016.

McMullan, R., *Environmental Science in Building*. Palgrave Macmillan, 7th edition, 2017.

McPherson, M. J., *Subsurface Ventilation and Environmental Engineering*. Springer, 2012.

Meiners, R. E. and A. P. Morriss (eds.), *The Common Law and the Environment: Rethinking the Statutory Basis for Modern Environmental Law*. Lanham, MD: Rowman & Littlefield, 2000.

Menell, P. S. (ed.), *Environmental Law*. Ashgate Publishing Co., 2002.

Mihelcic, J. R. and J. B. Zimmerman ·(eds.), *Environmental Engineering: Fundamentals, Sustainability, Design*. John Wiley & Sons, 2014.

Miller Jr., G. T. and S. E. Spoolman, *Living in the Environment Concepts, Connections, and Solutions*. Belmont, CA: Brooks/Cole, Cengage Learning, 16th edition, 2009.

Miller, G. T. and S. Spoolman, *Environmental Science*. Cengage Learning; 16th edition, 2018.

Mines, R. O., *Environmental Engineering: Principles and Practice*. Wiley-Blackwell, 2014.

Mitchell, B., *Resource and Environmental Management*. Routledge, 2nd edition, 2018.

Moeller, D. W., *Environmental Health*. Harvard University Press; 4th edition, 2011.

Monteith, J. L. and M. H. Unsworth, *Principles of Environmental Physics*. Elsevier, 4th edition, 2013.

Moran, E. F., *Environmental Social Science: Human-Environment Interactions and Sustainability*. John Wiley & Sons, 2010.

Morgera E., *Corporate Accountability in International Environmental Law*. Oxford University Press, 2009.

Nathanson, J. A., *Basic Environmental Technology*. Prentice Hall, 2000.

Natural Sciences. Plymouth: The Higher Education Academy Subject Centre for Geography, 2008.

Nemerow, N. L., F. J. Agardy, and J. A. Salvato (eds.), *Environmental Engineering* (3 Volume Set). John Wiley & Sons; 6th edition, 2009.

New Horizons in Environmental Economics Series. Cheltenham, UK: Edward Elgar. 2009.

Niessen, W. R., *Combustion and Incineration Processes: Applications in Environmental Engineering*. Boca Raton, FL: CRC Press, 3rd ed., 2002.

Nilsson, M. and K. Eckerberg (eds.), *Environmental Policy Integration in Practice: Shaping Institutions for Learning*. Taylor & Francis, 2009

Nobel, P. S., *Environmental Biology of Agaves and Cacti*. Cambridge University Press, 2003.

Okereke, C., *Global Justice and Neoliberal Environmental Governance: Ethics, Sustainable Development and International Co-operation*. New York: Routledge, 2007.

O'Riordan, T. (ed.), *Environmental Science for Environmental Management*. Routledge, 2014.

O'Riordan, T. (ed.), *Environmental Science for Environmental Management*. Taylor & Francis, 2nd edition, 2014.

Palmer, J. and P. Neal, *The Handbook of Environmental Education*. Routledge, 2003.

Palmer, J., *Environmental Education in the 21st Century: Theory, Practice, Progress and Promise*. Routledge, 2002.

Pani, B., *Textbook of Environmental Chemistry*. I K International Publishing House, 2017.

Park, A., *Everybody's Movement: Environmental Justice and Climate Change*. Washington DC: Environmental Support Center, 2009.

Parker L. and K. Prabawa-Sear, *Environmental Education in Indonesia: Creating Responsible Citizens in the Global South?* London, UK: Routledge, 2019.

Pavithran, K. V., *A Textbook of Environmental Economics*. New Age International, 2008.

Pellow D. N., *What is Critical Environmental Justice?* Polity. Cambridge, 2018.

Pellow, D. N., *Resisting Global Toxics: Transnational Movements for Environmental Justice*. Cambridge: MIT Press, 2007.

Percival, R. V. et al. (eds), *Global Environmental Law at a Crossroads*. Cheltenham: Edward Elgar Publishing, 2014.

Percival, R. V., J. Lin, and W. Piermattei, *Global Environmental Law at a Crossroads*. Edward Elgar Publishing, 2014.

Perelet, R. et al., *Dictionary of Environmental Economics*. London, UK: Routledge, 2001.

Perman, R. et al., *Natural Resource and Environmental Economics*. Pearson Education, 3rd edition, 2003.

Peters, C. A., *Environmental Engineering Science*. Mary Ann Liebert, Inc., Publishers, 2019.

Pfafflin, J. R. and E. N. Ziegler (eds.), *Encyclopedia of Environmental Science and Engineering*. Boca Raton, FL: CRC Press, 5th edition, 2006

Phaneuf D. J. and T. Requate, *A Course in Environmental Economics: Theory, Policy, and Practice*. Cambridge, UK: Cambridge University Press, 2017.

Pinn D. E. (ed.), *Environmental Education: Perspectives, Challenges and Opportunities*. Nova, 2017.

Plater, Z. J. B. et al., *Environmental Law and Policy: Nature, Law, and Society*. Aspen Publishers, 4th edition, 2010.

Purdom, P. (ed.), *Environmental Health*. Academic Press, 2nd edition, 2013.

R. N. Stavins et al., (eds.), *Economics of the Environment: Selected Readings*. Edward Elgar, 7th edition, 2019

Rahman, M. H. and A. Al-Muyeed, *Water and Environmental Engineering*. Dhaka, Bangladesh: ITN-BUET, 2012.

Ramkumar, M., K. Kumaraswamy, and R. Mohanraj (eds.), *Environmental Management of River Basin Ecosystems*. Springer, 2015.

Ramsay, J. and J. Schroer, *Environmental Biology*. Kendall Hunt Publishing Co., 2020.

Rao, C. S., *Environmental Pollution Control Engineering*. New Delhi, India: New Age International Limited Publishers, 2nd edition, 2007.

Rathinasamy, M. et al. (eds.), *Water Resources and Environmental Engineering II: Climate and Environment*. Springer, 2019.

Rathore, H. S. and ?L. M. L. Nollet (eds.), *Pesticides: Evaluation of Environmental Pollution*. Boca Raton, FL: CRC Press, 2012.

Reible, D., *Fundamentals of Environmental Engineering*. Boca Raton, FL: CRC Press, 2019.

Reis, G. and J. Scott (eds.), *International Perspectives on the Theory and Practice of Environmental Education: A Reader*. Springer, 2018.

Reiss, M. and J. Chapman, *Environmental Biology.* Cambridge University Press, 2nd edition, chapter 1, 2000.

Revelle C. S., E. Whitlatch, and J. Wright, *Civil and Environmental Systems Engineering.* Pearson Education Limited, 2013.

Richardson B. J. and S. Wood (eds.), *Environmental Law for Sustainability: A Reader.* Hart Publishing, 2006.

Rieuwerts, J., *The Elements of Environmental Pollution.* London: Routledge, 2015.

Rinfret, S. and M. C. Pautz, *US Environmental Policy in Action.* Palmgrave Macmillan, 2019.

Robson, R. E. and D. Blake, *Physical Principles of Meteorology and Environmental Physics: Global, Synoptic and Micro Scales.* World Scientific Publishing, 2008.

Royal Society of Chemistry, *"Environment,"* https://www.rsc.org/campaigning-outreach/global-challenges/environment/

Russ A. and M. E. Krasny, *Urban Environmental Education Review.* Comstock Publishing Associates, 2017.

Russo, M., *Environmental Management: Readings and Cases.* SAGE Publications, 2nd edition, 2008.

Salmon, P. and D. Grinlinton (eds.), *Environmental Law in New Zealand.* Thomson Reuters, 2015.

Salvato, J. A., N. L. Nemerow, and F. J. Agardy, *Environmental Engineering.* Hoboken, NJ: John Wiley & Sons, 5th ed., 2003.

Salzman, J. and B. H. Thompson, *Environmental Law and Policy.* Foundation Press, 2003.

Sands, P. and J. Peel, *Principles of International Environmental Law.* Cambridge, UK: Cambridge University Press, 2012.

Sarkar D. et al. (eds.), *An Integrated Approach to Environmental Management.* John Wiley & Sons, 2015.

Saylan C. and D. Blumstein, *The Failure of Environmental Education (And How We Can Fix It).* University of California Press, 2011.

Schaltegger, S., R. Burritt, and H. Petersen, *An Introduction to Corporate Environmental Management: Striving for Sustainability.* Routledge, 2017.

Schlosberg, D., *Defining Environmental Justice: Theories, Movements, and Nature.* Oxford University Press, 2007.

Schwarz, P. M., *New Energy Economics Book.* Routledge, 2018.

Scotford, E., *Environmental Principles and the Evolution of Environmental Law.* Oxford, UK: Hart Publishing, 2017.

Shallcross, T. and J. Robinson (eds.), *Global Citizenship and Environmental Justice.* Editions Rodopi, 2006.

Sharma, P. D., *Environmental Biology and Toxicology.* Rastogi Publication, 2005.

Shastri, S. C., *Environmental Law in India.* Eastern Book Company, 2nd edition, 2005.

Shelton, D., A. C. Kiss, *Judicial Handbook on Environmental Law.* UNEP/Earthprint, 2005.

Shelton, D., *International Environmental Law.* Brill-Nijhoff; 3rd edition, 2004.

Shigeta, Y., *International Judicial Control of Environmental Protection; Standard Setting, Compliance Control, and The Development of International Environmental Law by the International Judiciary.* Kluwer Law International, 2010.

Shogren, J. F. (ed.), Experiments *in Environmental Economics.* Ashgate Publishing Co., 2003.

Shrader-Frechette, K., *Environmental Justice: Creating Equality, Reclaiming Democracy.* Oxford University Press, 2002.

Singh, H. R., *Environmental Biology.* S. Chand Publishing, 2nd edition, 2004.

Singh, P. and T. A. Wani, *Basic Environmental Physics.* Pragati Prakashan, 2016.

Singh, P., A. Kumar, and A. Borthakur (eds.), *Abatement of Environmental Pollutants: Trends and Strategies.* Elsevier, 2019.

Singh, Y. K., *Environmental Science.* New Delhi, India: New Age International, 2006.

Small, J. C., *Geomechanics in Soil, Rocks, and Environmental Engineering.* Boca Raton, FL: CRC Press, 2016.

Smith, C., *Environmental Physics.* New York: Routledge, 2004.

Smith, P., *"Environmental health,"* in J. O. Nriagu (ed.), Policies. Oxford University Press, 2002.

Spaargaren, G., A. P. J. Mol, and F. H. Buttel (eds.), *Governing Environmental Flows: Global Challenges to Social Theory.* Cambridge, MA: The MIT Press, 2006.

Spellman, F. R., *Handbook of Environmental Engineering.* Boca Raton, FL: CRC Press, 2016.

Spellman, F. R., *The Science of Environmental Pollution.* Boca Raton, FL: CRC Press, 3rd edition, 2017.

Spooner, A. M., *Environmental Science for Dummies.* Hoboken, NJ: John Wiley & Sons, 2012.

Squires, G., *Urban and Environmental Economics: An Introduction.* Routledge, 2012.

Steady, F. C., *Environmental Justice in the New Millennium: Global Perspectives on Race, Ethnicity, and Human Rights.* New York: Palgrave Macmillan, 2009.

Sterner, T. (ed.), *The Economics of Environmental Policy: Behavioral and Political Dimensions.* Edward Elgar Publishing, 2016.

Stevenson, R. B. et al. (eds.), *International Handbook of Research on Environmental Education.* Routledge, 2012.

Stewart, A., *Developing Place-responsive Pedagogy in Outdoor Environmental Education.* Springer, 2020.

Sudani, B. R., *Analytical Environmental Chemistry.* New Delhi, India: AkiNik Publications, 2018.

Sullivan, J. B. and G. R. Krieger, *Clinical Environmental Health and Toxic Exposures.* Linppcott Williams & Wilkins, 2001.

Tang, W. Z. and M. Sillanpää, *Sustainable Environmental Engineering.* John Wiley & Sons, 2018.

Thampapillai, J. and J. A. Sinden. *Environmental Economics - Concepts, Methods and Policies.* OUP Australia & New Zealand, 2nd edition, 2013.

The National Research Council, *Global Sources of Local Pollution: An Assessment of Long-Range Transport of Key Air Pollutants to and from the United States.* National Academies Press. 2010.

Theodore, L. and R. R. Dupont, *Environmental Health and Hazard Risk Assessment: Principles and Calculations.* Boca Raton, FL: CRC Press, 2012.

Therivel, R. and B. F. D. Barrett, *Environmental Policy and Impact Assessment in Japan.* Routledge, 2020.

Tickner, J. A. (ed.), *Precaution, Environmental Science and Preventive Public Policy.* Washington: Island Press, 2002.

Tietenberg, T. and L. Lewis, *Environmental & Natural Resource Economics.* Pearson Education, 9th edition, 2009.

Tietenberg, T. and L. Lewis, *Environmental Economics & Policy: Global Edition.* Pearson, 6th edition, 2013.

Tietenberg, T. H. and L. Lewis, *Environmental Economics and Policy.* Addison-Wesley Professional, 2009.

Tietenberg, T. H. and L. Lewis, *Environmental Economics: The Essentials.* Taylor & Francis, 2019.

Tripathi, G., *Modern Trends in Environmental Biology.* CBS Publishers and Distributors, 2002.

Trivedi, P. R., *Environmental Pollution and Control.* Ashish Publishing House, 2004.

Tykva, R. and D. Berg (eds.), *Man-Made and Natural Radioactivity in Environmental Pollution and Radiochronology.* Springer, 2004.

Vaccari, D. A., P. F. Strom, and J. E. Alleman, *Environmental Biology for Engineers and Scientists.* Wiley-Interscience, 2005.

Vaccaro, I., E. A. Smith, and S. Aswani (eds.), *Environmental Social Sciences: Methods and Research Design.* Cambridge University Press 2010.

van Tatenhove, J., B. Arts and P Leroy (efs.), *Political Modernisation and The Environment: The Renewal of Environmental Policy Arrangements.* Springer, 2013.

Vanloon G. W. and S. J. Duffer, *Environmental Chemistry - A Global Perspective.* Oxford University Press, 2000.

Vesilind A. and T. D. DiStefano, *Controlling Environmental Pollution: An Introduction to the Technologies, History, and Ethics.* DEStech Publications, 2006.

Vesilind P. A., J. J. Peirce, and R. F. Weiner, *Environmental Engineering.* Elsevier Science, 3rd edition, 2013.

Vig, N. J. and M. E. Kraft, *Environmental Policy: New Directions for the Twenty-First Century.* Sage Publicatons, 10th edition, 2019.

Walker, G., *Environmental Justice: Concepts, Evidence and Politics.* London, UK: Routledge, 2012.

Wehrmeyer, W., Greening People: *Human Resources and Environmental Management.* Routledge, 2017.

Weidner, H. and M. Jänicke (eds.), *Capacity Building in National Environmental Policy: A Comparative Study of 17 Countries.* Berlin, Germany: Springer Verlag, 2002.

Weiner, E. R., *Applications of Environmental Chemistry: A Practical Guide for Environmental Professionals.* Boca Raton, FL: Lewis Publishers, 2010.

Weiner, R. and R. Matthews (eds.), *Environmental Engineering.* Butterworth-Heinemann, 4th Edition, 2003.

Weiner, R. E. and R. A. Matthews, *Environmental Engineering*. 2003, Burlington, M: Elsevier Science, 4th edition, 2003.

Weiner, R., R. Matthews, and P. A. Vesilind, *Environmental Engineering*. Elsevier Science, 2003.

Welford, R. (ed.), *Corporate Environmental Management 1: Systems and Strategies*. Routledge, 2nd edition, 2016.

Wiesmeth, H., *Environmental Economics: Theory and Policy in Equilibrium*. Springer 2012.

Williams, L. D., *Environmental Science Demystified*. New York: McGraw-Hill, 2005.

Wisner B. and J. Adams (eds.), *Environmental Health in Emergencies and Disasters: A Practical Guide*. World Health Organization, 2002.

Wolfrum, R. and N. Matz, *Conflicts in International Environmental Law*. Springer Science & Business Media, 2003.

Woodhouse, J. L. and C Knapp (eds.), *Place-Based Curriculum and Instruction: Outdoor and Environmental Education Approaches*. Psychology Press, 2000.

Wright R. T. and D. Boorse, *Environmental Science: Toward a Sustainable Future*. Pearson; 13th edition, 2016.

Wright, J., *Environmental Chemistry*. Taylor & Francis, 2005.

Wulfhorst, J. D. and A.K. Haugestad (eds.), *Building Sustainable Communities: Environmental Justice and Global Citizenship*. Editions Rodopi, 2006.

Yassi, A. et al., *Basic Environmental Health*. Oxford University Press, 2001.

Zaheed, I. H., *A Text Book on Environmental Biology*. New Delhi, India: Discovery Publishing House, 2013.

Zehnder C. et al., *Introduction to Environmental Science*. University System of Georgia, 2nd Edition, 2018.

Zhang, J. Y. and M. Barr, *Green Politics in China: Environmental Governance and State-society Relations*. Pluto Press, 2013.

Zhao, J. et al. (eds.), *Rock Mechanics in Civil and Environmental Engineering*. Boca Raton, FL: CRC Press, 2010.

INDEX

Printed in the United States
by Baker & Taylor Publisher Services